SOUND RECORDING

SOUND RECORDING

THE LIFE STORY OF A TECHNOLOGY

David L. Morton Jr.

The Johns Hopkins University Press
Baltimore

Printed in the United States of America on acid-free paper

First published in 2004 by Greenwood Press

Johns Hopkins Paperbacks edition published 2006

9 8 7 6 5 4 3 2 1

The Johns Hopkins University Press
2715 North Charles Street
Baltimore, Maryland 21218-4363
www.press.jhu.edu

ISBN 0-8018-8398-9 (pbk.: alk. paper)

Library of Congress Control Number: 2005936765

A catalog record for this book is available from the British Library.

Published in an edited paperback edition by arrangement with Greenwood Publishing Group, Inc., Westport, CT.

To Nan

Contents

Introduction

The story of sound recording is in some ways a history of history. Just like language, the visual arts, writing, printing, and photography, sound recording is a way for people to capture and store the events of the present and revisit them in the future. Yet with the exception of photography, which is only a little older than sound recording, all these methods of keeping records have been in use for centuries. Our ability to capture sounds, rather than simply describe them, came late in the nineteenth century, and since that time we have recognized the muteness of the great majority of human history.

While we are now able to capture our aural history, we still do not often take advantage of that ability. If one could gather the whole body of recordings that exist today into a single historical collection, it would be clear almost immediately that the vast bulk of the collection contains only music. This historical fact makes it difficult to write about the history of sound recording without focusing on music and the music industry. Perhaps in the future there will be more preserved recordings of voices or ordinary sounds, but for now, the story of recording is dominated by the story of *music* recording. Fortunately, music is a profoundly important part of culture and history, and an excellent historical resource. This book surveys the history of the devices used to record that musical cross section of history, showing how music, the music business, and recording technology have co-evolved. However, readers who are more interested in music than technology may be disappointed by the emphasis on machines and the ways they were used. While this is not primarily a history of music, it is also true that there have been many crucial junctures where music and musicians shaped the development of recording technology, and the most important of those are discussed here. But the content of the recordings themselves is the subject of these chapters only in passing.

Another goal of this book is to draw together the many threads of the history of sound recording, weave them into a coherent narrative, and present that story in a way that is accessible to a broad readership. Scholarship in the field does not do a good enough job of presenting the "big picture," nor is it written for the general reader. The present work addresses the need for a survey history that is based on solid scholarship, but is aimed at an audience beyond the small circle of academics. Toward that aim, the present volume is intended to be a reference work and guide to a history that cannot possibly be captured entirely in a single volume. Further, if it breaks any new scholarly ground, it is as a synthesis rather than as the document-based analysis that is typical of academic history today. I thus owe a great debt to the many scholars whose fine work I have mined. In taking their best ideas and data and reworking them for a new audience, I hope I have not done them a disservice. I have labored to give them due credit in the notes and bibliography.

The book begins with the story of the earliest experimental recordings, made to study the nature of sound waves, and concludes with developments and controversies that have emerged since the turn of the twenty-first century. With such a broad chronological sweep and just seventeen chapters to work with, it was necessary in some places to condense, but hopefully without introducing serious distortions. Wherever possible, the scope is international, but the available sources and my own limitations have necessarily made this a very "American" study.

Those wishing to probe further into the history of sound recording should start with the works cited in the notes, which point toward the best available historical literature, and which in turn may lead readers toward the sources of original historical documents and artifacts. Every effort has been made to include only readily available, published or online sources, although in some cases those sources are now out of print or are only available in specialized libraries. Also, I have cited references to "primary" or archival resources sparingly, because of my belief that the general reader will not have easy access to these sources, and because so much of the archival data on early recording history is treated so masterfully by the authors of the secondary literature.

Timeline

1857 Leon Scott in France demonstrates the phonautograph system for recording sounds.

1877 In April, Charles Cros writes a description of a machine that records sound much as a phonautograph does. He proposes using a chemical process to transfer the record to a permanent medium that will allow it to be reproduced. He apparently never demonstrated a working prototype.

1877 In July, Thomas Edison files a patent on sound recording and reproduction.

1878 Tinfoil phonographs are made by several small machine shops at Edison's request. These are distributed to demonstrate the principle of the phonograph.

1886 Chichester Bell and C. S. Tainter patent important improvements on Edison's original phonograph. They call their machine the graphophone.

1888 Inventor Oberlin Smith publishes the first article on magnetic recording technology.

1889 In January, the Columbia Phonograph Co. begins its commercial life, based partly on the patents of the earlier Graphophone Co.

1894 Emile Berliner introduces the gramophone in the United States (1889 in Europe), using a disc instead of a cylinder and a groove cut from side to side (laterally cut) instead of Edison's "hill and dale" (vertically cut) method. The gramophone is aimed at the entertainment market, and home versions are not capable of making recordings.

1899 Thomas Edison demonstrates his motion picture device with sound based on the phonograph.

1900 Danish inventor Valdemar Poulsen demonstrates a practical magnetic wire recorder at the Paris Exposition.

1901 Victor Talking Machine Company formed from a merger of Emile Berliner's company and the manufacturing works owned by Eldridge Johnson.

1906 Victor Talking Machine Company offers its first Victrola, a disc phonograph.

1912 Following years of declining cylinder sales, Edison at long last begins offering disc phonographs and records for sale.

1924 In October, Columbia Records and Victor Talking Machine Co. experiment with "electrical" recording equipment developed by Western Electric. Electrical recording employs electronic amplifiers, microphones, and electromagnetic record cutters.

1929 Edison ceases production of records and pulls out of the home phonograph business. Thomas A. Edison, Inc.; continues to be a major force in the dictation equipment business and continues to use cylinders through 1950.

1933 First demonstration of the AEG/I. G. Farben magnetophon tape recorder.

1946 Demonstrations of the magnetophon in the United States, Great Britain, and elsewhere.

1948 Commercial introduction of tape recorders in the United States.

1948 Columbia Records introduces the 33⅓ rpm long-playing (LP) record disc.

1949 RCA Victor introduces its 45 rpm disc and a special record changer on which to play them.

1950 RCA Victor issues its first recordings on the Columbia LP format.

1951 Columbia issues its first recordings on the RCA 45 rpm format.

1954 The transistor radio is introduced.

1957 In December, the Recording Industry Association of America formally adopts the Western Electric standard for stereo disc recordings.

1958 RCA-Victor introduces its first stereo LPs and tapes.

1962 European introduction of the Phillips compact cassette. The cassette is introduced in the United States in 1964 under the Norelco brand name.

1965 Stereo 8 (or 8-track) tape player system is introduced.

1979 Sony introduces the Walkman.

1982 Phillips begins marketing its laser-read compact disc player.

1986 Sales of LP records decline to 110 million units; tapes sell more than three times as many units, compared to 50 million units for the compact disc.

1987 The digital audio tape (DAT) system is announced, sparking an ongoing controversy among lawmakers and record companies.

1988 CD sales top LP sales for the first time.

1992 MPEG-3 (later called MP3) announced.

1

Birth of Recording

BACKGROUND

Thomas Edison's 1876 invention of the phonograph is fondly remembered as one of the technological milestones of the late nineteenth century. The phonograph was the machine that launched the record industry. While the phonograph has been memorialized as the world's first sound-recording device, that is simply not true. Edison, the beneficiary of centuries of thought on the nature of sound, drew his inspiration from a line of scientific instruments that had originated with the ancient Greeks. Physicists today know that sound travels through the air (or water, or walls, or any medium) in the form of waves. These waves occur very rapidly, so that it is probably easier to think of them as vibrations. Humans and many other land-dwelling creatures have evolved sensitive hearing, perhaps reflecting the fact that the air is a relatively poor transmitter of sound waves compared to, say, water. Nonetheless, our ears detect sound waves or vibrations across a wide range of frequencies, and our brains perceive them as sounds.

The study of sound, a field of science now known as acoustics, began around the sixth century B.C.E. when the Greek philosopher Pythagoras studied music and musical instruments. He was interested in the properties of stringed instruments, in which the strings visibly vibrate when they are plucked to produce sounds. In the early modern period, sound was one of the natural phenomena that fascinated physicists. Galileo, for example, discovered the principle of resonance, which he called sympathetic vibration. In the early 1600s, French mathematician Marin Mersenne described the vibration of strings mathematically, leading to formulas known today as Mersenne's Laws. Such attempts to describe sound waves mathematically would lead to general theories of waves, which had applicability to the whole spectrum of wave phenomenon including light and electromagnetism.

Experiments with sound waves grew more common in the eighteenth and nineteenth centuries, as new laboratory devices were invented to measure or study many aspects of sound vibrations. For example, Jean-Marie Constant Duhamel, a French scientist, discovered that by strapping a pen to an ordinary tuning fork and then tapping the fork, the pen would scribe a wiggly line on a piece of paper. For the first time, it was possible to see what a sound wave might look like. An English physicist, Thomas Young, in 1806 applied Duhamel's apparatus to a rotating cylinder coated with wax, in an experiment that would presage later versions of the phonograph.

THE PHONAUTOGRAPH: ALMOST A PHONOGRAPH

A variation of this instrument, invented in 1856 by Frenchman Leon Scott, was called the phonautograph. This device used a cone-shaped horn to capture sound and "focus" it on a flexible membrane stretched across the small end. Sounds captured by the horn made the membrane vibrate rapidly. Linked to the membrane through a delicate mechanism was a pointed stylus, to which the vibrations were transmitted; it too vibrated rapidly. Scott mounted a smooth glass cylinder on the mechanism so that the stylus would lightly touch the surface of the cylinder. Before use, the cylinder would be held over a flame to give it a dark coat of soot. When all was ready, someone could shout or play a musical instrument near the horn; sound vibrations would be transmitted to the stylus, which would begin to vibrate. If the cylinder was turned rapidly during this, the stylus would scribe a thin line in the soot, rendering a visible record of the sound. This was most likely the first true instance of a sound being recorded.

Later versions of the phonautograph used flat pieces of glass, discs, or even strips of paper, but the basic principle was the same. The device was copied and used in many laboratories and classrooms in the United States and Europe during the 1860s and afterward. There is an old rumor that Scott visited Washington, D.C., to demonstrate the phonautograph, and that he recorded the voice of Abraham Lincoln, but there is no known evidence to support this and none of the recordings have been found. Joseph Henry of the Smithsonian Institution also purchased a phonautograph in 1866 and used it for experiments and demonstrations, and he might have recorded famous voices, although apparently none of these have survived, either.

In fact, original phonautograph recordings of any kind are extremely rare. A few of them were copied and reproduced in scientific textbooks, but many of them, especially those recorded in soot on glass, were simply discarded. However, the surviving traces from the 1850s and 1860s are the earliest sound recordings of any kind. Several people have suggested that sounds could be reproduced from these recordings using modern technology. It has been possible for many years to translate sounds into visible patterns, such as when sound recordings are made on motion picture film to create the soundtracks for movies exhibited in theaters. Using a technology similar to that used to reproduce such "optical" recordings, it should be possible to reproduce visible phonautograph recordings and listen to these early recordings for the first time. However, the recordings are such crude representations of the original sound waves that it is unlikely that it would be possible to distinguish words. In any event, a phonautograph recording was usually just a moment of sound—a syllable of a word or a single note played on a violin. The resulting recordings would not be very interesting to hear.

An experiment conducted by the History Center of the Institute of Electrical and Electronics Engineers (IEEE) in 2000 was intended to put an end to speculation about the reproduction of phonautograph recordings using modern digital signal processing (DSP) techniques. One of the top DSP researchers in the United States volunteered his university laboratory for the experiments, and this author set about finding phonautograph samples. A few well-preserved phonautograph traces were discovered among the holdings of the archive of the Thomas Edison National Historic Site in West Orange, New Jersey. National Park Service staff made high-quality digital scans of the traces, but the original recordings were so badly made that there was little to try to recover, and the experiment was abandoned.

ALEXANDER BELL AND THE EAR PHONAUTOGRAPH

The most surprising thing about the IEEE effort was the realization that phonautograph recordings are today so rare, since the phonautograph was such a well-known piece of scientific equipment. It is a standard feature of textbooks on acoustics from the late nineteenth century, and examples of these devices are preserved in museums around the world. One phonautograph experimenter was none other than Alexander Graham Bell, inventor of the telephone in 1876. Before his telephone days, Bell was best known as a researcher in the field of sound and deafness. In the early 1870s he was living in Boston and working as a professor at Boston University. He witnessed demonstrations at Massachusetts Institute of Technology (MIT) of two methods of visualizing sound: an improved phonautograph developed by Charles Morey, a student of the famous MIT professor Charles Cross, and the "manometric flame" of Rudolph Koenig, which used sound waves to alter the shape of a gas flame, making it possible to "see" sounds. Bell was intrigued by the fact that patterns of sounds recorded by the phonautograph were significantly different than the patterns created by the same sounds when seen through the manometric flame. He believed that the devices were not capturing the sounds accurately, and began thinking of ways to improve them. At his summer workshop in Brantford, Ontario, Canada, he tested a very special phonautograph of his own design in an effort to capture sound waves accurately, and ultimately to gain a better understanding of human hearing. Instead of an ordinary horn and diaphragm, as in the Scott device, Bell used the ear, part of the skull, and the internal hearing-related parts of a cadaver obtained for him by a colleague. Bell used these actual body parts to construct what he called the "ear phonautograph" in 1874. Softening the dead tissue by rubbing it with glycerin and water, Bell found that the ear's internal mechanism would become pliable enough to respond to sound. He attached a hay stalk to one of the tiny ear bones, and arranged the glass-sliding mechanism of a Morey phonautograph so that the hay would scratch out a trace on the glass. The ear itself, Bell reasoned, was the most accurate mechanism for recording sound.

THE "FRENCH EDISON"

The phonautograph was clearly capable of recording sound, but what remained was for someone to seek a way to reproduce those recordings. Given the popularity of the phonautograph, it is not too surprising that more than one person conceived of a sound recorder capable of reproduction. What historians refer to as simultaneous invention, when two (or more) inventors working independently arrive at the same idea at nearly the same time, is quite a common phenomenon in the history of technology. Particularly in the nineteenth and twentieth centuries, inventors around the world have addressed the same kinds of problems in the same ways. They patented, wrote about, or lectured on their ideas; influenced each other with their ideas; and consciously or unconsciously borrowed concepts and made them their own. The inventors who are remembered in history are occasionally not those who invented first, but those who succeeded in making their inventions known to the public. In April 1877, just months before Edison patented the phonograph, a Frenchman named Charles Cros (not to be confused with Charles Cross of MIT) disclosed his ideas about a sound recording and reproducing device based on the phonautograph, which he called the parleóphon. Not having the funds to apply for a French patent, he instead wrote up a description of the machine and, as was common then in France, sent it to the Académie des Sciences. What Cros proposed was to make an ordinary phonautograph recording in the form of a spiral on a disc. The graphite-coated disc would then be used to make a permanent record of the recording, using a well-known photo-etching process. In effect, this process worked by using acid to etch the delicate undulations into the surface of a metal plate. The acid would attack the areas where the stylus had scraped away the carbon, but the remaining areas would remain untouched. If the etched plate were then replaced on the parleóphon, the groove would now reverse the recording process, driving the stylus and diaphragm back and forth. The original sound, he reasoned, would then be reproduced. In an October 1877 issue of the magazine *La Semaine du Clergé*, the writer Abbé Lenoir described Cros's invention and even gave it the alternate name "phonograph."

EDISON AND THE PHONOGRAPH

Thomas Edison clearly knew of some of these earlier experiments. He certainly knew of the phonautograph, but probably not the parleóphon. What directed Edison's attention was not the study of sound for its own sake, but rather Alexander Bell's recent invention of the telephone. Bell announced the telephone to the public in 1876, and Edison, already a famous inventor in his own right, took note immediately. At the time, Edison was working on several different projects in the field of telegraphy, the area in which he had established his reputation in the 1870s.

Virtually all of Edison's inventions up to that point had been built around the technology of the telegraph, but he was about to make a major departure. Born in 1847, he had grown up around the railroads and worked his way into a job as a telegraph operator as a teenager. After becoming more familiar with the inner workings of the telegraph system, he made his first invention in 1867 at age twenty. This device used incoming telegraph signals to power a paper-embossing device, creating a record of the incoming telegraph message on a strip of paper tape. That paper tape could then be fed into a second device, which "read" the indentations and transmitted a copy of the original telegraph message. Edison reasoned that recording at normal speed but transmitting at a much higher speed would result in messages taking much less time to travel down the wires. The growing telegraph companies had a perpetual shortage of what today would be called "bandwidth." Telegraph traffic had steadily increased in the nineteenth century, but for many years it was only possible to send one message at a time on the wire. Because human operators had to tap out the Morse code messages by hand, they could only go so fast. Edison's machine might have helped pack more messages into the wires by speeding things up considerably. The invention was not entirely novel—others had conceived virtually the same machine, and he never sold one of the devices—but it set Edison down an inventive path he would follow for the rest of his life.

He moved to Boston in 1868 to work for the telegraph company Western Union, and continued to invent there. That year, he patented a device he called the Vote Recorder, which was intended to register votes automatically and hence improve the speed and accuracy of voting. While it too fell by the wayside, it encouraged Edison to turn his full attention to a career as an inventor, which he soon began to do. Over a period of several years in the late 1860s and 1870s, Edison worked in two shops in New York City and in Newark, New Jersey, producing improvements to telegraph technology and trying to interest investors in sponsoring his work or purchasing his inventions. Two of the inventions that emerged during this period, the improved stock ticker and the quadruplex telegraph, earned Edison a reputation as an expert in telegraphy. The stock ticker was based on earlier devices that printed out incoming Morse telegraph signals on a paper tape, but it printed them in plain letters. That made it possible for people without training in Morse code to read the messages. Its biggest application was, as the name implied, installations in the offices of Wall Street stockbrokers. Today's "ticker tape parades" in New York no longer use real ticker tape, but the giant heaps of scrap paper tape that used to clutter the streets after one of these parades were an indication of how successful this invention was. The second invention was important to telegraph service providers. Unlike his earlier telegraph tape embosser, which simply sped up messages, the quadruplex telegraph used special electrical circuits to stuff up to four messages on a single wire at the same time. Edison's device was merely an improvement over another inventor's duplex (two-message) telegraph, but it was an important breakthrough and resulted in considerable commercial success. As so often happens in the history of technology, experiments with one technology led to the invention of something to serve a very different purpose. Edison became interested in the use of some kind of recording device to capture the output of a telephone and possibly retransmit it later.

TELEPHONE TECHNOLOGY

Because the telephone was the basis of so many phonograph-related inventions later, it is important to understand how it works. Its basic principle, in today's engineering terms, is called "transduction," the conversion of acoustic waves in the air to waves of electromagnetism or pulses of electrical current. Electromagnetic waves can travel through space or can be channeled along a wire, just as sound waves travel through the air or can be channeled inside a pipe or along a string. Without getting too far into the theory of electrical circuits, suffice it to say that Bell's original telephone generated electrical waves in a circuit by using the minute undulations of the air waves, acting on a sensitive mechanical device, to vary or "modulate" the flow of electrons in a circuit. In addition to modulating a current, a second way to convert sound waves to electric waves was to generate a new current through electromagnetic transduction. A magnet moving past a wire generates a tiny current in the wire as it passes. Using sound waves to move a tiny magnet will produce a tiny, but usable, current flow in a wire placed nearby. There are other ways to accomplish the same end, but these two methods of the transduction of sound constituted two of the earliest forms of the microphone. Now, the microphone is an essential part of any telephone. A telephone consists of a microphone, or transmitter as it was then known, linked by wires to a "receiver," a device that is a microphone in reverse, transducing electrical flow into sound. The electromagnetic type of microphone described above functions equally well as a telephone receiver. The rapidly undulating current emanating from a microphone, if passed to a second such device, causes the second device to reproduce sound as the incoming current in the wire drives the magnet back and forth. The resulting vibrations set up sound waves in the air. And although the purpose of telephone receivers and transmitters is very different than that of the telegraph, the batteries, wires, and other parts of the telephone were taken almost directly from telegraph technology. That is why telegraph expert Thomas Edison took such an immediate interest in Bell's telephone.

EDISON'S TELEPHONE RECORDER

Although Edison arrived too late to invent the telephone, he set to work immediately to improve it. One of his first ideas was a telephone recording device. In one early proposal, Edison sketched a machine that employed a small, flexible disc with a sharp point mounted at its center. This assembly was suspended above a strip of paper coated with wax, with the tip of the point just touching the wax. Shouting at the disc would cause it to vibrate microscopically, so that if the paper was pulled under the point, it would record the movement of the diaphragm, and therefore it would record the sounds that caused that movement. This in itself was no breakthrough, but it reflects Edison's earlier telegraph work and also was the beginning of Edison's effort to record the telephone in a similar way. A later proposal was to attach the point directly to the receiver of a telephone. In the Bell telephone receiver, incoming electrical signals flowed through a coil of wire, causing a small diaphragm to vibrate, and it was this vibrating diaphragm that generated the sounds. A telephone receiver-stylus vibrating under the influence of an incoming electrical signal, held against a strip of waxed paper pulled under the device, resulted in the wax receiving a record of the vibrations in the form of a string of indentations. Edison suggested that the indentations might then be used to reverse the process. Pulling the recorded tape under the modified receiver would allow the indentations to drive the stylus, setting up a current that could be transmitted along wires, which would recreate the original sounds through an ordinary telephone receiver connected to the wires. First mentioned in his laboratory notes in July 1877, the idea sat idle for several months until Edison returned to it that November.

Edison discarded the idea of recording the telephone, probably after he discovered through experiments that the recordings were too weak or distorted to be heard clearly through the telephone receiver. How he came to change his mind is not clear from his laboratory notes, but by November 1877 he had stopped working with telephone receivers and concentrated on recording sounds directly from the air, just as the phonautograph did.

The Edison sound recorder was prematurely announced in the United States by Edward Johnson, one of Edison's talented associates, in the November 17, 1877, issue of *Scientific American*. Although at this time a working machine had probably still not been constructed, Johnson's description closely resembled some of the ideas recorded in Edison's laboratory notebooks of that period. Now the device's telegraphic and telephonic origins were hard to recognize, as the machine did not require electricity in any way. Johnson described a machine where the user would talk into a speaking tube (rather than the phonautograph's horn), and the resulting sounds

Edison's original tinfoil phonograph, constructed in late 1877. The model shown here does not have a sheet of tinfoil in place. U.S. Department of the Interior, National Park Service, Edison National Historic Site, West Orange, New Jersey.

acted on a sensitive diaphragm at the end of the tube. Attached to the diaphragm was a stylus or "embossing chisel" that was forced up and down by the sound waves. The point of the chisel recorded the motion on a strip of paper similar to that used before and in fact not so different than the paper tapes used for years in telegraph printers. Reproduction was the reverse of this, except that Johnson still claimed that if the diaphragm of the reproducer was built into a telephone transmitter, the record could be transmitted as a telephone signal. The reality was that Edison had in July experimented with a device that electromagnetically embossed telephone signals onto a wax-coated paper tape, but he was far from a practical device and would soon drop the whole idea of telephone recording.

It may be surprising from today's perspective to learn that perfecting such a monumental invention as the phonograph was not the sole focus of Edison's attention at this time. Over the period from the late summer of 1877 to November of that year, Edison flitted between projects, returning only occasionally to the phonograph. But by November, his laboratory notebooks show that he had changed his approach to the invention. Now it consisted of a hand-cranked, rotating cylinder, wrapped with a sheet of tinfoil instead of waxed paper. The tinfoil he used was perhaps three or four times as thick as today's kitchen aluminum foil, so it was thick enough to stand up to the stylus but flexible enough to be indented using only the power of acoustic energy. The cylinder had a shallow, spiral groove machined into its surface, which provided some space into which the tinfoil was pushed as it was embossed. The recording/reproducing stylus was attached not to a telephone receiver but to a simple diaphragm, which in turn was stretched over the small end of a funnel-shaped horn just a few inches long. In this respect, it was very similar to the Scott phonautograph. Edison's machinist, John Kruesi, constructed a prototype from Edison's drawings during the first six days in December 1877. Edison would later recount that he had shouted the first few verses of "Mary Had a Little Lamb" to make the first recording on the prototype, and was then amazed that the device actually worked the first time he tried it. He would write that "I was never so taken back in my life." The phonograph was born.[1]

1. In a 1906 account, Edison recited the invention of the phonograph, including his recollection that the first words he uttered were "Mary had a little lamb, etc."

2

Out of the Laboratory

THE MEANING OF SUCCESS

The commercial success of sound recording was never inevitable. Throughout history, inventions have overcome technical obstacles without subsequently becoming successful. The patent records are full of useful inventions, but only a small fraction of them are ever manufactured, are used by large numbers of people, or come to have a broader social significance. Yet the phonograph would be one of those rare inventions that did succeed. Part of the reason was that Edison had the resources to see his invention through to commercialization. For sound recording to succeed, it needed not only the phonograph itself but also a champion who would manufacture, distribute, and promote it or find others who would.

In December 1877, Edison had submitted a U.S. patent application and had announced the phonograph to the public, but was still working with the cantankerous tinfoil recorder. Despite the fact that it had worked the first time he tried it, subsequently it proved to be notoriously difficult to operate, and the resulting recordings were often badly garbled. It took a high degree of skill to get consistent results from the recorder, a fact that convinced Edison of the need for immediate improvements. That did not stop him from dreaming of how the masses would use the invention. He wrote notes to himself proclaiming that the likely applications of the phonograph would include toys, talking clocks, talking advertisements, and music recorded on flat sheets. The delicate tinfoil recordings could be copied to a much sturdier metal medium by an electrotyping or molding process, similar to processes used in other industries and, in the case of electrotyping, similar to the process suggested by Cros. These metal copies could, Edison promised, be played indefinitely. Further, at other times Edison mentioned using permanent metal copies to make (probably by stamping) large quantities of duplicates for distribution. Edison thus dimly foresaw a multitude of commercial applications for the invention, including some that anticipated the future of musical entertainment.

A MASTER OF PUBLICITY

Edison set out both to improve the phonograph and to promote it; he saw both of these activities as essential, and both contributed to the phonograph's wildly enthusiastic public reception. Thomas Edison is remembered as one of the greatest inventors of all time, and in his day was sometimes called "The Wizard of Menlo Park," referring to the laboratory he had set up in rural Menlo Park, New Jersey, at about this time. While Edison was already well known in the telegraph industries of the United States and Europe by the late 1870s, it was the phonograph that catapulted his reputation to the status of a popular hero. Following his earlier inventions, such as the stock ticker, the quadruplex telegraph, and others, Edison had used his own and his investors' money to set up what he would call an "invention factory" in New Jersey in late 1875. At Menlo Park, he and his workers could pursue their experiments in relative peace and quiet, away from the distractions of Newark and New York City. But the laboratory was also conveniently located within walking distance of the main railway that ran from New York City to Philadelphia, so that necessary trips to the cities were convenient and quick. It was from this base of operations that Edison set about transforming the phonograph from a laboratory gadget to a household item, yet the first stage was to demonstrate the still-crude device under carefully choreographed conditions.

With the construction of the first tinfoil phonograph in December 1877 and its successful demonstration in the laboratory, Edison made the phonograph ready for the public. Almost immediately, he and two associates took the recorder and caught the train from Menlo Park to lower Manhattan. They marched to the offices of *Scientific American* magazine, today a monthly featuring mainly science, but in that day a weekly featuring some science but a great many articles on inventions and new machines. Its excellent reputation and wide readership made it the perfect place to announce the phonograph, as Edison was sure that the editors would take it upon themselves to write up an article about the demonstration for immediate publication.

Perhaps because the phonograph recording process was still so difficult to accomplish, Edison had prepared a cylinder with a recording on it. In the enthusiastic review of the demonstration published some days later, *Scientific American*'s editor gushed that "the machine inquired as to our health, asked how we liked the phonograph, informed us that *it* was very well, and bid us a cordial good night. These remarks were not only perfectly audible to ourselves, but to a dozen or more persons gathered around" (*Scientific American*, 1877, 384).

The reaction could not have been more favorable. The trickle of reporters making their way to Edison's invention factory grew to a steady stream, and it was just a few months later that the New York *Graphic* newspaper featured a well-illustrated feature story on the inventor, his lab, and his phonograph, and dubbed him Wizard of Menlo Park. Edison, who had already predicted that his phonograph would be used in talking dolls, clocks, and other contraptions, now began additional statements to the press about its future commercialization. In one famous quote, he stated in his trademark vernacular that the phonograph "is my baby and I expect it to grow up to be a big feller and support me in my old age." In June 1878, in an article on the phonograph that appeared in the popular magazine *North American Review*, Edison told the public how the phonograph would be useful; the first step toward marketing such a novel invention was not to convince people that they should want one, but simply to explain what it was and what it could do. It could be used, he said, for sending recorded business letters that did not have to be transferred to paper (sort of a precursor to today's voice mail), for making recordings of loved ones and famous figures for future generations to enjoy, for creating talking books, for teaching speech and music, for advertisements, for talking toys and clocks, and (although he had already tried it and failed) for recording the telephone. Edison sought to be not merely the inventor of the device but also the inventor of the public's uses of it, and hopefully their desire for it.

COMMERCIALIZING THE PHONOGRAPH

Edison set about turning his invention into income almost immediately. In 1878, he was offered $10,000 by Gardiner Hubbard, an investor who also

In April 1878, Edison was invited to Washington, D.C., to demonstrate his phonograph to the National Academy of Sciences. He stopped by the studios of the famous photographer Matthew Brady, who took this picture of the thirty-one-year-old inventor with an early model of the machine. U.S. Department of the Interior, National Park Service, Edison National Historic Site, West Orange, New Jersey.

happened to be involved in the newly formed Bell Telephone Company of Boston. In return for the payment, Edison agreed to transfer the manufacturing and sales rights to the new firm. This company became the Edison Speaking Phonograph Company, the first firm organized to commercialize sound recording. The company at first made up a batch of about thirty machines, intending to sell them at a shop in New York City. However, the machinist who made them apparently was unable to duplicate Edison's laboratory model, and the resulting machines did not work well.

As an alternative, Edison worked to design a machine reliable enough to be taken on the road for demonstration purposes. In the 1870s, a recording of sound was such a novelty that people would pay money to hear a demonstration of it. Edison collected thousands of dollars in royalties during the first year of exhibitions, although this income stream rapidly diminished in later years. James Redpath, who ran an organization that sponsored public speeches and demonstrations, was hired to organize regional representatives to tour with the phonograph. Redpath's men were trained in the art of making recordings and were given one phonograph each, along with a supply of tinfoil recording blanks. Then they toured the country, putting on shows for local audiences and splitting the admission fees with the company.

The Edison Speaking Phonograph Company refused to sell its machines outright, preferring instead to lease them to the licensed exhibitors. That is in part because the machines were so difficult to use and so unreliable that the company's leaders believed that putting them in unskilled hands would result in public outrage. Putting the first generation of machines in the hands of the public would, they thought, result in instant dissatisfaction with the invention, and the resulting publicity would damage the reputations of both the phonograph and Edison. They hoped that the next generation of improved phonographs would overcome these difficulties and that when they became available, the originals could be called in to be destroyed.

During 1878, Edison and his assistants worked on improvements aimed at making the recording process simple and reliable. They failed. Instead, the operation of the machine continued to demand a practiced hand, and even then sometimes it simply refused to record sounds well enough for them to be intelligible when they were played back. Perhaps it was fortunate that demand for the demonstration phonographs began to decline by 1879, when the novelty wore off. In late 1878, Edison turned to other projects, such as improvements to the telephone and, more significantly, the creation of his famous electric lighting system.

In turning away from an invention that had so captivated the public (if only briefly), Edison had probably made a tactical blunder. Admittedly, the electric lighting system that emerged from his research around 1882 proved to be a huge commercial success (spawning what is now the General Electric Company and the predecessors of many of today's electric utility companies). He clearly intended to return to the phonograph someday, but in the meantime his neglect allowed other inventors to leapfrog the basic phonograph technology, making it necessary for Edison to try to catch up later.

THE INVENTION OF THE GRAPHOPHONE

The process by which other inventors entered the field began to occur during and after 1881. Alexander Graham Bell had won a major award, the Volta Prize, for his invention of the telephone. He used the prize money to set up a new experimental research laboratory in Washington, D.C., which he called the Volta Laboratory. There, Bell employed his oddly named cousin, Chichester Bell, and a talented machinist with the equally unlikely name of Charles Sumner Tainter. Bell and Tainter somehow obtained one of Edison's early tinfoil phonographs (or made a copy) and substituted soft wax for the tinfoil recording medium. The wax was coated onto a metal cylinder or, later, a cardboard tube that could be slipped on and off a permanently mounted metal cylinder (which became known as the mandrel). Because the resulting recording was weak, small rubber tubes carried the sound directly to the listener's ears. A recording was made by speaking into a large funnel that resembled the one first used on the Scott phonautograph, rather than Edison's small speaking tube.

Tainter's laboratory notes, which are today preserved in the archive of the Smithsonian Institution's National Museum of American History, detail a long process of experimentation and the development of several novel means of recording sounds. Late in 1881, the group deposited a box at the Smithsonian Institution containing their latest wax cylinder recorder. A cylinder was included in the box, along with a card indicating that the following words were recorded on the wax: "I am a graphophone and my mother was a phonograph." But it was not until 1885 that the Volta team announced its wax cylinder recorder to the public.

The 1885 graphophone had been modified somewhat in design since its original incarnation in 1881, when it was literally a copy of an early Edison phonograph with wax pressed into the cylinder grooves. The newer machine used a longer, wax-coated cardboard cylinder with dimensions of about six inches in length and just over an inch in diameter. Like later Edison tinfoil models, the graphophone had separate recorder and reproducer assemblies. Listening tubes were supplied that ran directly to the ears from the reproducer. The machine was still like the early phonograph in other respects, such as the need to supply its motive power by hand cranking.

Gardiner Hubbard was recruited to present the graphophone to the leaders of the Edison Speaking Phonograph Company in an attempt to interest them in an investment in the new invention. The Edison Company delayed, so the Volta group in Washington decided to form their own company in 1886, which was incorporated in nearby Alexandria, Virginia, as the Volta Graphophone Company. This company at once started marketing the graphophone for entertainment purposes. But a demonstration of the machine attracted the attention of Andrew Devine, who worked as a "reporter" or secretary in the U.S. Supreme Court. Devine became an investor in the company and brought in an acquaintance named James O. Clephane, who was also a court reporter. Other demonstrations brought interest from John H. White, a reporter for the U.S. House of Representatives. In part because of the enthusiasm of these men for a machine that could "take notes" with perfect accuracy, Volta Graphophone turned to marketing the device for court reporting and other forms of note taking. The Graphophone Company began to play up the technical features of their machine that were well suited to note taking. The graphophone cylinder length, for example, meant that the recording was spread out over a broader area than the phonograph, making it easier for a user to locate a particular passage on the cylinder and move the reproducer to it when necessary. The cylinders themselves could be easily placed into or lifted out of the machine, unlike the tinfoil medium that had to be wrapped onto the phonograph and that could easily be torn. In its commercial form, the graphophone used a treadle mechanism like a sewing machine, which provided a simple but fairly steady rotary motion. Although this feature was not directly related to its use for taking dictation or notes, it was a familiar and reliable source of power, having been used on sewing machines for many years.

EDISON'S RESPONSE: PERFECTED PHONOGRAPH

Edison may never have acknowledged it, but the appearance of the graphophone clearly sounded an alarm that sent him scrambling to produce his own improved sound recorder. During late 1886 and into 1887, Edison, his assistant Charles Batchelor, and others worked to duplicate the wax-cylinder graphophone and add a few patentable improvements. At first they concentrated on the wax recording medium. The graphophone's wax was so soft that it tended to dislodge and collect on the recording stylus, resulting in degraded performance. While the tinfoil phonograph had embossed its recording rather than incising it, Edison seems to have accepted incising for wax recordings. He and Charles Batchelor experimented with a number of chemical compounds, trying to find a slightly harder one that would flake off and scatter as it was carved out of the groove during recording. They discovered a suitable compound when they combined paraffin wax with a natural resin. Edison also experimented with more sensitive reproducers that would result in louder reproduction, and new materials for recording and reproducing styli. Being an expert in electricity, he also designed a small electric motor and battery for the phonograph. By late 1887, Edison announced a wax-cylinder phonograph with an electric motor and other new features, but it was still evident that the new machine was not yet finished.

In early 1888, Edison and his team continued a diverse range of research activities that encompassed more than just the basic recorder and medium. One of his staff, Arthur Kennelly, was given responsibility for designing a new electric motor and battery for use with a phonograph. Another staff member, Jonas Aylesworth, experimented with other new wax compounds. A third, Theo Wangemann, made experimental musical and voice recordings in an effort to improve recording techniques, and still another expert, Franz Schulze-Berg, researched ways to duplicate recordings for commercial production.

The same momentum that had led Edison toward a graphophone-like design for the new phonograph also tended to steer him toward thinking of his invention more as a dictation device. As the final design began to take shape, it was clear that Edison had office dictation in mind rather than entertainment or the other applications he had suggested in the 1870s. Like the graphophone, Edison's new phonograph had a mechanism to allow rapid starts and stops, so that passages on the cylinder could be repeated if necessary. Another feature added to increase its appeal as a dictating machine was the shaving attachment, which would let the user shave the surface of the wax and allow cylinders to be reused several times. The cylinders themselves were no longer coated onto cardboard, but consisted of a solid tube of a new material, discovered in 1888, which was chemically similar to soap. Finally, in the summer of 1888, Edison announced the "perfected phonograph" with great aplomb. Just as he had done in 1877, Edison posed for a photograph with the new machine, except that in this photograph he looked more exhausted than triumphant. Nonetheless, the picture marked his return to the phonograph field.

Edison's investors established a new firm, the Edison Phonograph Company, and had Edison set up a new factory next to the laboratory in West Orange to replace the earlier phonograph manufacturing plant at Bloomfield, New Jersey. Meanwhile, Edison's intention to enter the dictation machine business to compete with the Graphophone Company did not prevent him from simultaneously pursuing other phonographic applications. In 1887, inventors William Jacques and Lowell Briggs had shown Edison an idea for a talking doll using a tiny record player placed inside the doll's body. In 1888, Edison assigned Charles Batchelor the duty of improving the design, and formed a company with the two inventors to manufacture and distribute it. Dolls and tiny record players were made for

In the summer of 1888, Edison again posed for a photograph, this time with his new "perfected" phonograph. Edison, now forty-one, had nearly exhausted himself in the rush to introduce the recorder. From left: Fred Ott, W. K. L. Dickson, Charles Batchelor, Edison, A. Theodore Wangemann, John Ott, Charles Brown, and George Gouraud. U.S. Department of the Interior, National Park Service, Edison National Historic Site, West Orange, New Jersey.

the Edison Toy Phonograph Company at the new phonograph factory in West Orange. But, plagued by technical problems in building the delicate, miniature record player, the company missed its goal of supplying dolls to stores for the 1889 Christmas season. When the dolls finally went into production in early 1890, they were still so unreliable that Edison decided to stop selling them after only 3,000 had been produced. The company ceased operations altogether in October 1890 and Edison abandoned the idea of a talking doll (it would be revived by others in later years and would become one of the most popular children's toys of all time).

As the production of dictating machines was getting underway in 1888 and 1889, Edison was looking ahead to a second entertainment-related application. He proposed the creation of yet another new firm, which would probably have been called the Edison Amusement Phonograph Company.

Looking back, his team's experiments on the mass duplication of recordings and the improvement of recording techniques had little to do with the business phonograph, but hinted at his interest in creating a market for the entertainment phonograph. However, the entire phonograph and graphophone businesses would soon become so entangled in complex legal and business dealings that Edison found it difficult to follow up on this initiative. Commercialization of both devices would fall to a third party, a firm called the North American Phonograph Company, of which Edison was not yet aware.

3

The Commercial Debut of Sound Recording Devices

UNDECIDED

With the development of the new graphophone and "perfected" phonograph between about 1886 and early 1888, the Edison and Graphophone companies were ready to begin the process of marketing their inventions. Their task was to convince some segment of the public of the value of phonographs. Clearly they were able to do so, but how did it happen? Unfortunately, the early commercialization of the phonograph and graphophone business generated a number of lawsuits and complex business deals that make telling the story more difficult.

The manufacturing company Edison had set up in the 1870s to make tinfoil phonographs was simply dissolved in the 1880s, after the introduction of the perfected phonograph. The graphophone interests also reorganized, and the original Volta Graphophone Company dropped from the scene after it licensed its business to a new firm called the American Graphophone Company.

Then, beginning in the late 1880s, both phonographs and graphophones were marketed by the North American Phonograph Company, which also helped establish numerous local phonograph companies around the United States to service and sell sound recorders. North American Phonograph had purchased the marketing and distribution rights (in an underhanded way, according to Edison) to both inventions, effectively wresting control from the inventors. Outside of the United States, there were several other companies making and selling sound recorders by the 1890s, some of them still directly affiliated with Edison or the graphophone investors.

North American Phonograph's founders adopted the approach to marketing that was first articulated by the leaders of Volta Graphophone. Their belief was that the main market for recorders would emerge from the ranks of professionals engaged in fields that required careful note taking. While the leaders of Volta Graphophone had been congressional reporters and used the invention in that work, there were also thousands of courts of law where sound recording could also be used. Because the graphophone had the lead in this market by late 1887 or 1888, Edison adapted his phonograph to this purpose, making design changes that were suited to the needs, as he perceived them, of note-takers. North American Phonograph customers were apparently offered the choice of either of the machines, so in some sense the two still competed. In fact, because some of the first customers shied away from the battery-powered phonograph in favor of the treadle-powered graphophone, Edison hastily introduced a foot-powered phonograph in order to keep abreast of the competition. Improvements aimed at making the phonograph superior to the graphophone sometimes backfired. For example, the introduction of shaving machines allowed Edison's wax cylinders to be reused, but this innovation created problems in other areas. Edison's machinists were forced to redesign critical parts when they discovered that the recording/reproducing mechanism of the phonograph had to be painstakingly adjusted to account for the varying diameters of shaved cylinders. Graphophones, which had single-use, wax-coated cardboard cylinders that could not be shaved, did not require such an adjustment or a redesign of the machine.

Despite the improving technology, the effort to find customers for the new dictation machines, whether phonograph or graphophone, was a nearly complete failure. The recorders were still too difficult for novices to use, and the idea of "mechanical stenography" was just not yet catching on. Only about 3,000 machines, most of them phonographs rather than graphophones, were in service by 1891. Sound recording was starting to look like a failed technology.

However, change was afoot. Edison had never really put the idea of an entertainment phonograph out of his mind, and in early 1889 the Edison Phonograph Works manufactured a selection of recorded musical cylinders, sending them to North American Phonograph to be distributed to the local companies. Intended for demonstration purposes, the cylinders were instead put into service for public amusement.

Several sources date the dawn of the modern recording industry to November 23, 1889, when Louis Glass, president of the small phonograph distributor serving California and the West Coast of the United States, installed his first coin-operated phonograph. It was based on the battery-operated, electric phonograph then being manufactured by the Edison Phonograph Works. The machine, installed in the Palais Royal Saloon in San Francisco, was an instant hit and began generating money for the cash-strapped company. In addition to adding the coin-handling mechanism, Glass adapted the machine to employ four sets of hearing tubes with a coin slot corresponding to each set of tubes. Within a year, Glass and others had arranged for the manufacture of mechanisms to convert the Edison phonograph to coin operation. Phonograph-operating companies around the country were astounded at the public's response to these early devices, despite the fact that they regularly failed to work and under the best of circumstances were only capable of playing a single recording.

Placed mainly in bars, but later also put into places where polite society could find them, these phonographs were soon generating more profits for some of the local phonograph companies than rentals of dictation recorders. Edison, who had by now developed ways to make multiple copies of recorded wax cylinders, latched onto the idea of turning record copying into a new business for himself. Unfortunately, because of his relationship with North American Phonograph, he was not yet in a good position to pursue that market.

EDISON'S RETURN TO LEADERSHIP

North American Phonograph's narrow focus on office dictation equipment and its overestimation of the market nearly led to its bankruptcy as early as 1891. Its local distributors had not fared well, either. That year, only nineteen of the original thirty-three local phonograph companies were still solvent, surviving only because the "side business" of providing entertainment was proving to be such a good moneymaker. The market would get even worse beginning in 1893 as the United States entered a serious downturn in the business cycle. North American Phonograph's weakness provided the opportunity for Edison to buy back a stake in the company, and he gradually assumed control over it during 1891 and 1892. Edison, by about 1893, had in effect regained his legal right to sell phonographs, although by now many of the original patents were set to expire and the market was very weak. He created yet another firm, the National Phonograph Company, to make both business and entertainment phonographs and to duplicate entertainment cylinders. Interestingly, he still assumed that the local phonograph companies (and others) would make the recordings, then send them to him in New Jersey, and that he would then duplicate them, although that was not the way it would happen in future years.

The making of entertainment recordings came to assume a role as important as the making of recorders and players. In 1891, the New York Phonograph Company (one of the local phonograph distributors) had purchased $15,000 worth of recordings, and presumably other companies were spending almost as much. Edison set up a recording studio on the third floor of his laboratory to make the master recordings, and engineer Walter Miller made improvements to the special recording phonograph used in the studio. A new production-model phonograph, designed previously in 1889, allowed variations in the construction of the basic mechanism so that different versions of the phonograph could be tailored for either home or office use. This machine in entertainment form, designated the Type M, sold for $150 to $200 beginning in 1893. Further adaptations for the home phonograph included the abandonment of the battery-powered electric motor used for dictation, substituting instead a spring-wound, clockwork motor. This lowered the price of the total package, required less maintenance, and reduced the weight of the machine. By 1896 additional improvements were introduced, and the standard model eliminated the recording attachment entirely, making it available only as an extra-cost option. The machine now sold for as little as $100.

The graphophone, meanwhile, was emerging from a state of inactivity following the failure of the dictation business. The original investors tried and failed to gain control of the bankrupt North American Phonograph Company, spurring Edison to take full control of the firm in order to dissolve and finally be rid of it. But one of North American's former licensees, the Columbia Phonograph Company in Washington, D.C., had acquired the right to make graphophones. This company now pushed forward on its own, adopting the phonograph-style cylinders for the graphophone and introducing a spring-wound home record player selling at a price less than that of the phonograph. Edison went back to the drawing board and soon had redesigned the Type M once again to produce the "Home Phonograph," which was even simpler than its predecessor and sold for only $40. Columbia retaliated with a $25 model, a challenge to which Edison responded with a $20 machine. By 1900, with the cost of players falling to ever-lower levels and the number of recorded cylinders rising, the phonograph-graphophone (at this point their essential features

This carefully posed picture from around 1900 was intended to convey the joys of mechanical entertainment. The two unidentified men pose with an early record player and listen through rubber tubes. U.S. Department of the Interior, National Park Service, Edison National Historic Site, West Orange, New Jersey.

The scale of the booming phonograph industry of the early 1920s is suggested by this image of the Columbia Phonograph Company's factory in Bridgeport, Connecticut. U.S. Department of the Interior, National Park Service, Edison National Historic Site, West Orange, New Jersey.

were virtually the same, so it is more convenient to refer to it simply as the "phonograph") had been thoroughly transformed into a mass-market product.

HOW THE EARLY ENTERTAINMENT RECORDINGS WERE MADE

In the early 1890s, when the first entertainment records were being made, the Edison Phonograph Works and the local phonograph distributors made their own records, duplicated them, and used them in coin-in-the-slot machines or sold them to the public. Edison had greatly improved the recording attachments of the home phonograph, but from the very start, it was clear that making a good entertainment recording would not be as simple as setting a performer in front of the recording horn. Unless performers stood very close to the horn and played or sang loudly, the resulting recording would be far too weak to overcome the phonograph's inherently high level of background noise. On the other hand, too loud a sound could result in distortion (this was especially true after the introduction of disc records). Certain types of instruments simply could not be recorded easily, while others, such as violins, recorded fairly well. Groups of performers had to crowd near the horn, which sometimes made it difficult to play. It became common later to put certain instruments such as the piano up on a platform to raise its soundboard closer to the horn. Often, the phonograph was located in an adjoining room, with the recording horn poking through a window or a hole in the wall. Making a recording was a delicate operation requiring a highly skilled "recordist." These technicians could tell by listening and watching how the recording session was going. They learned from experience how to substitute the diaphragms in the recording horn with membranes of greater or lesser flexibility, depending on the type of performance, the humidity of the air, or any of a host of other factors. These recordists were fairly well-paid individuals who often were jealously guarded by their employers. While Edison and others improved recording technology over the years, in general this is the way that recording sessions were conducted through the 1920s, when new technologies came along.

HOW RECORDS WERE DUPLICATED

Besides simply making the recording, the other key to developing an entertainment market was the invention of ways to duplicate recordings economically. Duplication of records was on Edison's mind shortly after the invention of the original tinfoil phonograph in 1877. Because these first records were metal, he imagined that he might lay the recorded tinfoil flat, make a plaster mold of it, and then press new sheets of tinfoil into the mold to make multiple copies of the record. However, with the advent of the wax cylinder graphophone and phonograph in the 1880s, this method was no longer possible. Now it was necessary to find some way to duplicate the fine indentations in a wax cylinder, which could not be flattened out like a piece of tinfoil.

Other late-nineteenth-century technologies provided a model for a way to make good duplicates of a wax recording, which was to make a mold of it and then cast copies in the mold. The promoters of the graphophone had given some thought to this in the 1880s, when they proposed electroplating the wax record (the same process used to apply a smooth, thin layer of chrome to an automobile bumper), stripping the plating off, and then using it to make a mold for making additional copies in wax. Edison applied for a patent on substantially the same idea in 1888, and in later years this method would be successfully employed to mass produce cylinders and discs.

In the meantime, though, many of the local record companies began to refuse to send Edison their cylinders, deciding instead to copy recordings themselves. Recording studios usually hired performers to appear before a bank of several phonographs, all operating simultaneously. That way, they captured several copies of a performance at the same time, so that the best one or two could be selected for duplication. Typically, a performance would also be repeated numerous times to produce multiple master copies from which to produce cylinders for sale. Then the best cylinders were copied. This was usually done by a "pantograph" technique. A special phonograph capable of holding two cylinders employed a delicate mechanism that linked the two. A recorded cylinder was played on one of the machine's mandrels and a blank placed on the other. The undulations in the groove of the recorded cylinder drove the mechanism, incising a copy of the recording into the blank groove of the second cylinder. About 25 to 100 copies could be made this way before the groove in the master record became too worn to produce a good copy.

Unfortunately, this method resulted in a significant loss of quality in the copied recording, but it was inexpensive and simple compared to the elaborate plating and molding that would have been required by Edison's technique. Ultimately, though, quality issues contributed to the pantograph duplication method's decline. Edison and other inventors, such as Thomas B. Lambert of Chicago, were making steady progress toward a practical and reliable molding technique that provided better sound and longer lasting records. Lambert actually beat Edison to the market in 1900 with cylinders molded in an unusual-looking pink celluloid plastic. Edison had improved his "gold molding" (plating) techniques by 1901 and introduced the first Gold Molded records that year. Molding soon became the standard for duplicating cylinders. There were many slight variations of the method, but in all of them the surface of a wax master cylinder was plated, either through the electrostatic deposition of gold on the surface or by rendering the surface conductive by coating it with powdered graphite and then electroplating the surface. The resulting metal copy, a sort of "negative" of the original, could be removed and built up by continued plating operations, then turned into a rugged mold. Injected into this mold was a melted mixture of materials; Edison used wax in the early years; a harder, synthetic wax beginning in 1908 (the "Amberol"); or a celluloid plastic ("Blue Amberol") by 1912. These materials shrank slightly when cooled, allowing the finished copy to be pulled from the mold. This became the standard way of duplicating cylinders through the end of the cylinder era.

The Lambert cylinders and Edison's later Blue Amberol cylinders introduced an important new material into the manufacture of records: plastic. Plastics in 1900 were already in use for buttons, combs, and various other small objects. Celluloid, probably the most common early plastic, could be heated into a soft mass and forced into a mold of any shape. When it cooled, it became harder than the earlier wax or soaplike compounds used earlier. Hardness was important, because in general a harder record produced more volume. That is because in the days of the early phonograph, the harder the stylus was pressed into the groove, the harder it drove the reproducer, and the more sound it produced in the horn. But pressing too hard quickly destroyed a soft wax recording, so plastic records represented a major quality improvement.

THE SCENE IN 1900

The phonograph was entering a new phase of its existence. What had once been a general-purpose technology for recording and reproducing was now splitting into two specialized applications. On the one side were the office dictation machines, which were seen as a nearly complete commercial failure at the time, but which would enjoy a return to prominence after 1900. On the other side were the entertainment phonographs. The Edison, Columbia, and other companies would continue to develop the entertainment market, which grew year by year. In 1890, it was the coin-slot phonograph that was generating the most income (mostly for its local operators).

Patrons of one of the early "phonograph arcades," probably around 1895. U.S. Department of the Interior, National Park Service, Edison National Historic Site, West Orange, New Jersey.

From the mid-1890s to about 1900, "phonograph parlors" were set up in many cities in the United States, where coin-operated phonographs were available to play a variety of entertaining records. Unlike the first installations, these were not in saloons or bars but public places where women and children were welcome. The idea also caught on in Europe, where for example the Pathé company (an Edison affiliate) employed forty people to run a highly popular phonograph parlor. The Edison and Columbia Graphophone companies were soon thriving, as were most of their European affiliates. However, the novelty of the phonograph would wear off and the importance of these public phonographs would fade. Meanwhile, Edison, Columbia, and others began to offer machines at prices that suited a mass market. In fact, most growth in the industry after 1900 came from the sale of home players and records. However, it was not the wax cylinder that would ultimately capture this market, but an upstart disc phonograph called the gramophone.

4

The Introduction of Discs

WHY DISCS?

The home record player was one of the great commercial success stories of the early twentieth century, yet that success was not all Edison's doing. Over the course of a decade from around 1900 to 1910, the pioneering technology of wax cylinders peaked in popularity and began to fade. In its place came the disc gramophone record, which would rise to dominance and remain as the standard form of sound recording for over seventy-five years. This type of disc remains in limited use to the present day. The transition from cylinders to discs in the early 1900s was less of a breakthrough in technology than a reflection of a market shift. Edison had thought about using discs in the 1870s, as had Charles Sumner Tainter and other experimenters at the Volta Laboratory. Both described disc players in their patents or laboratory notes, but then settled on the cylinder partly because it was their cylinder designs that they were able to make work well—the disc would require years of additional development before it could compete. The disc's success also hinged on a new set of champions. A new actor appeared on the scene in the person of Emile Berliner, an inventor who was previously known for an important invention in the telephone field. His system of disc recording and reproduction, introduced commercially

Emile Berliner as a young man, posed before mirrors, circa 1890s. Library of Congress, Motion Picture, Broadcasting and Recorded Sound Division, Washington, D.C.

around 1895, used many ideas taken from the cylinder phonographs already on the market but added a few twists.

Berliner believed that the main advantage of the disc was that it could be readily stamped out in a factory, perhaps more cheaply and easily than molded cylinder records. The new disc technology was from the start intended to be sold to consumers in the form of a record player, not a player/recorder like the phonograph or graphophone. Recordings would be made solely by manufacturers, not by consumers. Berliner perceived, correctly as it turned out, that his market would be customers who wanted

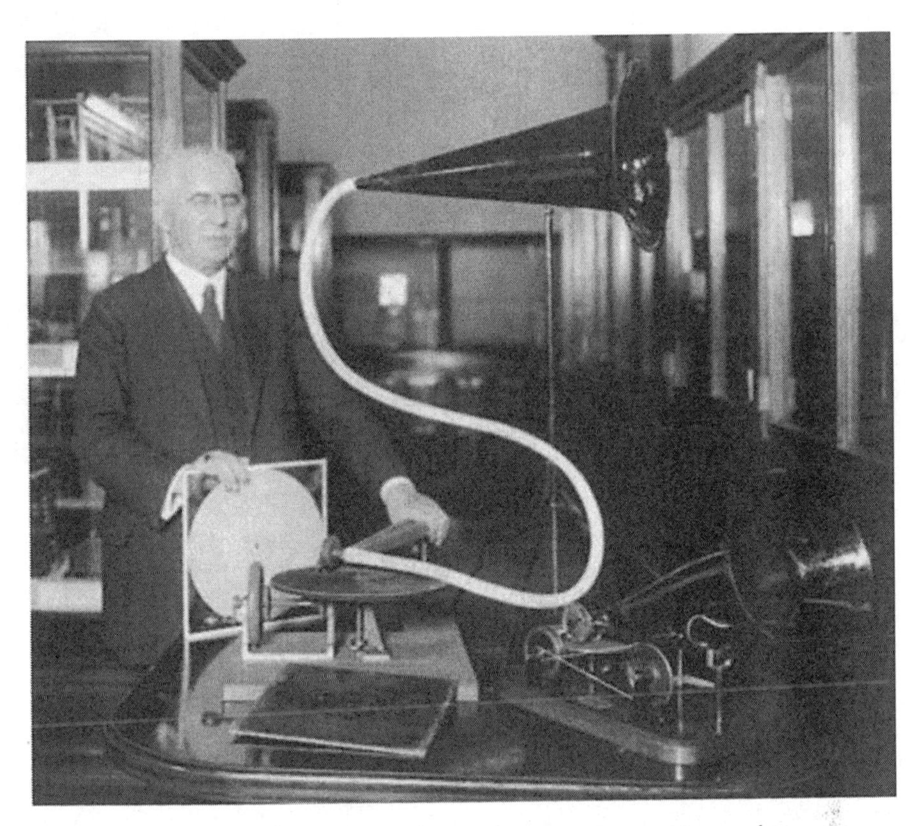

Emile Berliner in the 1920s with one of his early gramophones. Library of Congress, National Photo Company Collection, Washington, D.C.

to listen to recorded music in the home, but were not interested in making it themselves. Further, the new player as introduced in 1895 was cheaper than its competitors, although low-cost cylinder machines were forthcoming. Thus the gramophone was no major technical breakthrough, and the idea of a low-cost home machine to be used mainly for playing records hardly surprised the existing manufacturers.

In later years, disc gramophone players would come to be known as "phonographs" in the United States, even though that is technically incorrect. The English, Germans, and others insisted on calling them by their proper name. Yet the name was not Berliner's only lasting contribution. Within about a decade, sales of disc records and record players went from zero to the point where they dominated the market and drove the cylinder manufacturers out of the business or forced them to adopt discs. Even Thomas Edison felt compelled to answer the demand by introducing a disc system in 1912. It is therefore difficult to assign a simple explanation to the rise of the disc, yet it is necessary to try to explain its success.

BACKGROUND ON BERLINER

The disc's promoter, Emile Berliner, was a German immigrant working in America. He was born in Hanover, Germany, on May 20, 1851; had worked at a number of trades in his teens; and had made a few minor inventions as a young man. Then, in 1870, he sailed for America, settling first in Washington, D.C., and later in New York City. With thousands of others, he saw one of Alexander Graham Bell's telephone exhibitions in 1876. This fascinating invention spurred Berliner to begin building his own version, and that same year he developed an improved microphone for use in the telephone. He sold the patent for the microphone to the Bell Telephone Company, which used it (in modified form) for several years before replacing it with a newer technology (invented, ironically, by Edison). Berliner was nonetheless quite proud of this invention, and a sympathetic 1926 biography was subtitled "Maker of the Microphone." He was taken in by the Bell Company and worked on telephone improvements in New York and Boston in the early 1880s. By 1884, however, he was determined to establish himself as an independent inventor, so he left his post with Bell and returned to Washington, D.C. Then in 1886 he turned to the phonograph, perhaps after becoming aware of the work being done by the nearby Volta Laboratory. Although he left few records of his early experiments, he patented the gramophone in 1887. Early the next year, reading a paper on the machine before the Franklin Institute in Philadelphia, Berliner predicted that the future of sound recording lay in providing inexpensive recordings to consumers for home entertainment. It was a sentiment that had been expressed by others, notably Edison, but Berliner's technology would prove a better match to his intended market than either the phonograph or the graphophone.

THE BERLINER TECHNOLOGY

At the heart of Berliner's invention was not a new way of recording but a process for duplicating records. Although he tried several methods, he got the best results using a solid zinc disc, coated with a thin layer of wax. A special recording device scratched a record of the sound into the wax in a zigzag pattern, exposing the zinc below. The disc was then dunked in a chromic acid bath, which attacked the unprotected zinc wherever the recording stylus had carved away the wax coating. After a time, the acid created a shallow groove in the zinc disc. The etched disc could be played back if needed, which was a potentially valuable feature.

The etched disc was electroplated with a new layer of metal to make a "negative" copy for stamping. This new metal layer would be stripped off, backed up with many additional layers of plating to make it rigid, and then used to stamp out the records. The records were made by pressing a handful of a hard rubber compound, sold under the name Vulcanite, which was heated until it was softened. He would later substitute other plastic compounds for the hard rubber, and the mastering process would be improved. Like the disc idea, the notions of etching a master and then making molded or stamped records was hardly new. Both the Volta Laboratory researchers and Edison had demonstrated the idea in 1885 and 1891 respectively, and it bore a close resemblance to the ill-fated Cros recorder of 1877.[1]

There was nearly a seven-year lag between Berliner's first patents and the introduction of the first commercial disc records around 1895. During this period, the Edison and Columbia companies were beginning to gain momentum selling entertainment records. Berliner, while slow to start, benefited from the delay because others had developed the foundation of a consumer market, which he could then build upon. However, because the phonograph and graphophone had been gradually improved during that seven-year interval, the Berliner gramophone now seemed toylike in its simplicity, with its crude, hand-cranked design.

One advantage that the gramophone had over its competitors was volume. In the days of the acoustic phonograph, before electronic amplification, the maximum volume for record playback was directly proportional (in the case of cylinders) to the depth of the groove and how hard the stylus was pressed into the record during playback. Because the Berliner records were stamped from heated plastic, which hardened when it cooled, the records could stand a much greater stylus pressure before the groove began to succumb. The records also benefited from the process used to make the original master recordings in wax, which was so soft that it captured more of the original volume of the sounds fed to the recorder. However, this technical advantage was only temporary since Edison and others soon offered their own plastic records, and technical factors alone cannot explain the ongoing success of the disc.

1. U.S. Patent 372, 786, November 8, 1887.

INTRODUCING THE DISC COMMERCIALLY

Berliner, who had been making discs since 1893, began to sell them in 1895 or early 1896. Though his companies were seated in Philadelphia and Washington, D.C., some of his early commercial successes came from European sales. He was greatly aided by two former employees of Edison named Fred and Will Gaisberg. The brothers made trips to Europe in 1896 and later to attract "talent" and make recordings for mass duplication back in the United States. It was during one of these trips that the Gaisbergs made several important early recordings of European opera stars, notably Enrico Caruso. By recording these stars, the disc established an important commercial toehold in Europe, because opera and other "highbrow" forms of music were more avidly followed there than in the United States. That year the English Gramophone Company, Limited, was formed in England, which survives to this day under different corporate owners in the form of the HMV record label and the EMI Company.

Back in the United States, Berliner fitted a battery and electric motor to his phonograph in 1896, but this was not the form of the record player that was destined to gain the greatest success. In Philadelphia, a group of investors backed Berliner's effort to create a new firm, the United States Gramophone Company. Its first goal was to design a record player that would be cheap, reliable, easy to use, and better in quality than previous Berliner models. Fred Gaisberg would later recount that while investigating possible designers for a spring-wound motor for the gramophone, he found a machinist named Eldridge R. Johnson in a shop in nearby Camden, New Jersey. Johnson designed a new clockwork motor and, somewhat later, would make other improvements to the gramophone as well. Johnson would become increasingly important in the manufacture of gramophones, eventually taking over the business. While Berliner's first machines were simple and inexpensive, after 1905 many of the customers for the classical and opera recordings were wealthy, prompting Johnson to design a more elegant player that cost around $200 in 1906.

The early Berliner records, pressed into 7-inch diameter discs in hard rubber and selling for just 50 cents each, gave problems with warping and breakage, so a new plastic compound called Durinoid (used previously for making buttons) was substituted. Later, however, most records were made primarily from lacquer (an imported product derived, grotesquely, from insect excretions) and "lampblack" (powdered carbon), mixed with cotton flock (chopped cotton fibers) or ground rocks. The most important ingredients were the cotton or rock powder, which made up the bulk of

Eldridge R. Johnson. David Sarnoff Library, Princeton, New Jersey.

the record, and the lacquer, which held it all together. The lampblack was apparently used simply to color the records. The manufacturing process used through the 1950s, which was developed around the turn of the century, involved mixing and heating these ingredients, forming a ball of the mixture, and placing it into the stamper, which closed tightly to force the mixture into the grooves. The disc was then cooled and removed from the press. This formula would be modified somewhat, with Columbia and other manufacturers using cardboard discs coated with lacquer or plastic, and others substituting various fillers or eliminating the fillers entirely. The small discs were abandoned by 1906 or 1907, and 10 inches became the standard size. These held 3.5–4 minutes of music, up from the original 2 minutes.

Another technical innovation that was developed in those early years helped make the disc more appealing. While the technique of using a wax-coated zinc disc for making master recordings worked well, the harsh, acid-etching process resulted in too much background noise. The etching could not be easily controlled once set in motion, and it tended to eat into the zinc unevenly, resulting in random irregularities in the groove that translated into noise when a record pressed from this master was played. Between 1896 and 1898, Eldridge Johnson recalled that he developed an improved process for making master records, probably based on patented work by Edison. This involved the use of a solid wax disc for making a master recording. Once recorded upon, the wax master was removed from the recorder, brushed with fine graphite powder to make it conductive, and then electroplated in much the same fashion that Edison used for duplicating cylinders. The thin, plated layer or "mother" was then stripped off and additional plating layers added to beef it up, much as before. From this, a similar plating process was used to make a "master" record for archival purposes, as well as several stampers for the mass production of records. These plating processes were similar, but the solid wax record resulted in lower noise levels than the wax-coated zinc discs, and hence improved sound quality of the final product. With some changes, this process was still used to produce the small number of phonograph records still being made at the turn of the twenty-first century.

The gramophone was a great success in American cities like Philadelphia and New York, spurred on by the fashion precedent set by the phonograph and graphophone. By late 1898, Eldridge Johnson was producing parts to construct at least 600 players per week and was planning to double production. However, over the next two years the Berliner companies were embroiled in some complicated legal battles that resulted in Berliner being unable to sell records or players in the United States or use the "gramophone" trademark. Disc fever was nonetheless spreading, as the American Graphophone Company and other firms began offering their own discs and players (avoiding the Berliner patents through somewhat questionable means).

BIRTH OF VICTOR RECORDS

Eldridge Johnson, probably with the approval of Berliner, continued to make players and even moved into the recording and record-production ends of the business. In part to avoid the use of the Gramophone trademark, he registered the Victor brand in early 1901 and began putting it on his line of gramophone records. Soon, he would also adopt the corporate mascot registered in 1900 by Emile Berliner. The trademark was an image of a bull terrier gazing quizzically at the horn of a gramophone, accompanied by the words "His Master's Voice." Berliner had based the trademark on a painting by the English artist Jacques Barraud. This otherwise obscure artist had originally painted the dog looking into the horn of a cylinder player, but the artist revised the painting to make it a disc player after the work was sold to the English Gramophone Company, Limited. The little dog, "Nipper," would become one of the longest lasting and most recognizable of all trademarks, and he can still be seen in the corporate advertising of HMV (the initials come from the phrase "His Master's Voice") records and RCA electronics. This trademark was so successful that Berliner and Johnson made little effort to resuscitate the word "gramophone" after regaining the rights to use it, perhaps accounting for the fact that subsequently in the United States, the disc player eventually became known as the "phonograph." Between 1901 and 1903, Berliner sold his interest in the gramophone to Eldridge Johnson, who would rise to glory as leader of the Victor Talking Machine Company.

The Edison affiliates in Europe made some headway selling phonographs and cylinders, but the U.S. market was seen as having the greatest potential for growth. Although the Edison and Columbia companies, among others, had made some recordings of operatic and classical pieces, it was the comedy, novelty, and old familiar songs that sold the best. Yet Victor was about to make a major breakthrough with the release of a series of opera records. The Gaisbergs and Berliner, in Europe, and Johnson in the United States had made a number of recordings of European opera stars, and these were released in 1902 as a special series of Victor records with red labels, sold at a higher price. Despite the general consensus in the industry that "highbrow" music did not sell well, the recordings proved to be immensely popular in the United States, helping to launch the international career of tenor Enrico Caruso and inspiring a long-lasting series of Red Seal recordings for the Victor Company. It was Victor's coming of age.

CYLINDER MAKERS RESPOND

The success of the Berliner discs and players in the late 1890s was troubling to the cylinder makers. Edison used his extensive research facilities to stay at the forefront in developing new recording technologies, and had worked out improvements to the phonograph steadily over the years. One of his major advances was the introduction of larger Concert cylinders. This technology employed cylinders of five inches in diameter, which were considerably larger than the 3¾-inch diameter cylinders used since 1887. These Concert cylinders held more music than the earlier type, but cost $2.50 to $5.00 versus the 50 cents of the typical disc and took up the shelf space of a stack of about fifty discs. The cylinders also required the user to buy an expensive new player. Before even finding a way to mass produce these cylinders, Edison abandoned them around 1904 due to a lack of sales.

Even though Berliner and Eldridge Johnson were improving the disc, they were discovering its fundamental technical flaw. Some aspects of the sound quality of a phonograph recording can be improved by speeding up its rotation. That is part of the reason why Edison cylinders of the early 1900s turned at 160 rpm (the original cylinders of the 1890s turned at 120 rpm). In the gramophone disc system, which operated at about 80 rpm,[2] the "instantaneous speed" of the stylus in the groove changed dramatically as the record played. What that means can be understood easily by considering the beginning of a recording on a disc and the end of the same recording. At the beginning, the stylus is placed in the groove at the outermost edge of the disc. The groove is a spiral with turns that gradually get smaller near the center of the record. If the outside of the record is 10 inches in diameter, then it has a circumference of about 31 inches. If it makes 1 revolution per second, then the stylus travels over nearly 31 inches of groove per second at the beginning of the record. But near the end of the record the spiral is smaller, so the circumference of the last bit of the spiral may only be, say, 10 inches. Since the speed of rotation is still 1 revolution per second, then the stylus is traveling through the groove at a rate of just 10 inches per second. That is why the rate of the stylus in the groove is, in effect, constantly slowing as the record is played. And because a slower speed usually results in reduced sound quality, disc records had to be played fast enough so that they still sounded reasonably good as the stylus reached the inside grooves, but not so fast as to require an overly large disc diameter. Berliner's discs, in fact, simply were not recorded near the center of the disc, because near its center, the medium was simply unusable. This "slowing" was an inherent property of the disc, but it was entirely absent on a cylinder, where the spiral groove had a constant diameter.

Discs also represented a compromise in terms of size. Recordings lasting about 2 minutes (the same as a standard Edison cylinder) could be put on a 7–8-inch disc. Longer records soon appeared from Victor and others on 10-inch or 12-inch discs, and the smallest sizes were dropped. While Victor proposed using 14-inch discs, the size and fragility of the larger sizes

2. The 78 rpm speed was not standardized until after about 1912.

created problems. While a stack of 10- or 12-inch disc records was easier for consumers to store than a comparable collection of cylinders, in practical terms a 78 rpm disc would be limited to one short song per side.

The severity of these limitations did little to quell demand for gramophone records. Disc sales were taking off after 1904, sending the cylinder companies scrambling. Edison found that after 1900, his strongest market in the United States was rural. In the hinterlands, Edison's catalog of popular records (many of which seemed to appeal to traditional rural values) was better appreciated, and his reputation as an inventor and businessman gave consumers confidence in his products. Led by Victor's Red Seal label, urban buyers flocked to classical and operatic music on gramophone discs in ever-greater numbers. Price competition brought cylinder and disc players and records into line, but Victor also introduced some premium products that actually raised the price of some records. After abandoning the original 7-inch records in 1903 to offer 8-inch records at only 35 cents each, Victor also offered 10-inch and 12-inch discs at somewhat higher prices. ($1 and $1.50 respectively). By 1906, Victor would discontinue the inexpensive 8-inch records, as the sound quality was just too degraded at this small size.

As 10- and 12-inch discs came to dominate the market, more record companies began to abandon the cylinder. American Graphophone Company, one of the pioneer makers of cylinder recorders, was offering two models of a disc player that could play Victor-style discs by the early 1900s. Some companies were also imitating Victor's marketing tactics and catalog. Columbia, for example, which had introduced a successful line of popular and classical records under the Columbia label, mimicked Victor by offering a premium series of Grand Opera records that initially had red labels, just like Victor's Red Seal line. During these years, Victor was particularly successful in establishing its distribution and retailing system across the United States. By licensing wholesalers and allowing them to sell to established dealers (even cylinder dealers), Victor quickly built a distribution empire with twice as many outlets as its rival Columbia, which preferred to own all its local outlets. Competition increased as U.S. and European manufacturers began entering the market in greater numbers.

THE END OF THE CYLINDER

Discs had become so successful by 1910, particularly in the United States, that the end was clearly at hand for the cylinder. New companies, when they entered the record or player field, now usually adopted the disc format.

The Talk-O-Phone Company was one such entrant in 1904, and its success is suggested by company claims that it had sold 25,000 disc players in its first year. Victor scored a big hit in 1906 when it introduced a new player, the Victrola, with a horn that was hidden inside its wooden cabinet and thus out of view. The new record player was ideal for the middle-class parlor, where the phonograph's horn had always seemed out of place. The Victor Company earned $6 million that year, versus Columbia's $2.5 million. Few new manufacturers were entering the cylinder field in the years around 1910, when the cylinder was already headed into obsolescence. National Phonograph Company, which marketed Edison's products, closed its European branches by about 1909. Edison nonetheless introduced the expensive Grand Opera Amberola player that year, along with a new cylinder made of plastic, but sales of cylinder machines still fell. Columbia ceased production of cylinders in 1909, although it sold cylinders manufactured by a contractor until 1912. By 1913, Edison's last competitor in the cylinder market had also dropped out. Edison was left where he started, as the sole manufacturer of cylinder phonographs.

THE EDISON DIAMOND DISC

According to one account, it was Edison's employees who began designing a disc system, in secret, in 1910, only presenting the results to the "Old Man" later. Whatever the actual sequence of those events, it is clear that beginning in 1910 the Edison Company made its original recordings in both disc and cylinder formats. Following the company's reorganization as Thomas A. Edison, Incorporated, in 1910, Edison and his team finally moved forward toward the introduction of a new disc player. While the company continued to make a few cylinders for loyal customers through 1929, the symbolic death of the cylinder came in 1913 when the first Edison Diamond Disc players were announced to the public. At the same time as the triumph of the disc, sound recording was on the verge of its first commercial peak. The fifteen years after 1910 would see the record player become a widely diffused consumer item, one that helped define the soundscape of the early twentieth century.

5

Recording in the Business World

DICTATING A NEW ROLE FOR RECORDING

Before exploring the way that cylinder and disc phonographs became household items after 1910, it is worth returning to the history of what became known as the dictation machine. Thomas Edison in 1878 had predicted numerous scientific, educational, and entertainment uses for his phonograph, but several early attempts at commercialization failed, so he temporarily put the project aside. By letting it languish for some years while he pursued research on electric lighting, Edison opened the door for Alexander Graham Bell and his team at the Volta Laboratory in Washington, D.C., to carry forward his experiments. The resulting "graphophone" wax-cylinder recorder was to be made by the newly formed American Graphophone Company, a firm led by men who were keenly interested in marketing the device as a note-taking or stenography machine. Edison, who now returned to work on the phonograph, at first considered joining forces with the American Graphophone Company, but then backed out of negotiations and developed his own design, the graphophone-like "improved phonograph" of 1888. Edison now also focused on selling the phonograph for what he called mechanical stenography. Then, when the actual marketing and distribution of phonographs and graphophones were taken over by the North American Phonograph

Company, there was considerable momentum supporting the continuation of efforts to offer sound recorders for note-taking purposes.

HOW THE DICTATION MACHINE SURVIVED ITS CHILDHOOD

The original advertisements for the phonograph and graphophone claimed that note-taking work would be made more accurate if a recording were made rather than relying on handwritten notes. Then, by reviewing this recording, secretaries could make accurate transcriptions of what was said. At least at the beginning, this looked like a promising market, and in 1891 Columbia Phonograph (one of North American's local licensees) had about sixty machines leased out to customers around the District of Columbia, mostly federal government agencies, Congress, and the courts.

Yet neither the graphophone nor the phonograph succeeded in the later 1890s for stenographic purposes. The machines were so difficult to use that many early adopters soon returned them and canceled their leases. The regional distributors for the recorders, serving various territories around the United States, were sorely disappointed with the results and in turn drove the transition to the use of sound recordings for entertainment. In Europe, where the phonograph and graphophone were also marketed, the advent of mechanical stenography was hardly noticed except at a few overseas organizations run by Americans.

Yet mechanical stenography did not die. Instead, it expanded beyond government service and became a commonplace feature of business life in the United States and Canada. In the early 1900s, when the entertainment recording market was exploding, Edison practically withdrew from the field to concentrate on making records and improving the home phonograph. Yet he would return in 1905. Columbia Phonograph, which had become an independent entity and survived the disastrous early 1890s only by entering the entertainment market, also returned to mechanical stenography at about the same time. Part of the reason for this resurgence is that American government agencies and businesses were changing in the first decade of the twentieth century, leading to a new situation in which the phonograph could thrive in offices. Fifty years or so after the first truly large corporations had emerged in the railroad and steel industries, big business was in the throes of a management revolution. Whereas pen, paper, and hand labor had sufficed in the smaller, simpler companies of the nineteenth century, in the twentieth century a more mechanized form of clerical work had emerged that would, by about 1915, spawn the "scientific office management" movement (Hughes 1989).

THE SCIENTIFICALLY MANAGED OFFICE

Scientific office management emphasized the reorganization of office work using technologies (some of them quite simple) to increase efficiency. Clerks and other office workers were given more clearly defined duties and their power was compartmentalized into specific areas such as accounting, correspondence, and so on. Professional managers of the sort we have today were beginning to emerge. Following this managerial revolution were individuals and businesses eager to sell products and services to companies to help them make the transition.

Managers began to be bombarded with advertisements, catalogs, and trade publications that tried to influence their behavior or convince them to purchase some of the thousands of new business-related technologies that inventors developed in response to scientific management. The usual analogy was to the factory and the transition to standardized production in the nineteenth century. Just as factory managers had adopted both new practices and new technologies to create the Industrial Revolution, so too would office managers have to adopt both new ways of managing and new office technologies. Looking back, many of these may seem obvious, but at the time they were part of a major shift in the way business offices ran. Accounting practices, for example, were regularized, and new machines such as mechanical calculators were used instead of longhand arithmetic. Filing cabinets and file folders were introduced to replace the haphazard warren of desktop cubbyholes typical of the nineteenth-century businessman's desk. Typewriters replaced handwritten letters. Even the telephone's early success was also in part due to this movement. It was adopted by business users to increase the "efficiency" of communication between geographically separated offices in companies that were opening branch offices around the country. The new machines were more "scientific," because they promised to eliminate errors (Yates 1989).

Additionally, by 1900, managers in American corporations were doing to office work what Henry Ford would do to factory production. By dividing complex office procedures into smaller tasks and using machines for as many of those tasks as possible, managers could then hire low-skill employees to perform what had been high-skill work. Ford would prove how this worked on the shop floor, mass producing automobiles that the company could then sell at unprecedented low prices even though he paid his workers high wages. In offices, the "mechanization" of work not only changed the nature of the work but also affected who did it. Male workers, who had previously worked as high-skill, high-wage clerks or secretaries, were eliminated (although they could often move into management positions), and their work, now partly mechanized by typewriters and accounting machines, was taken over by women who were paid much lower wages.

These important shifts were underway at the same time that sound recording technology was emerging. In fact, when the first phonographs and graphophones were being sold in the late 1880s, there were still more male secretaries than female ones. But by about 1900, the situation was changing rapidly. As more women entered the workforce in the new, lower ranking clerical positions, often they found themselves working in mechanized, scientifically managed offices. In this environment, the new business phonographs began to sell more briskly and their manufacturers began tailoring their marketing campaigns to fit the new circumstances. Between about 1900 and 1905, the wording of advertisements shifted somewhat, to promote "business phonograph" and "mechanical dictation." Male users of recorders were portrayed "dictating" letters onto cylinders, which were then listened to by female "transcriptionists." While the basic form of the technology was the same, its uses had subtly changed.

EDIPHONES AND DICTAPHONES

Innovations in the stenographic phonograph and graphophone came faster after 1905, by which time the two technologies had become almost indistinguishable in their features. However, both the machines and the organizations that made and distributed them had become distinct from the entertainment phonograph branches, and so the "business phonograph" and graphophone became the focus of new divisions within the National Phonograph Company (Edison's firm) and the competing Columbia Phonograph Company. By 1906, both companies began to split off the office recorder divisions, with Edison creating a new commercial department and Columbia renaming its product the Dictaphone. Some years later, the Edison recorder would be renamed the Ediphone in response, so it is convenient to begin referring to them by these names rather than phonograph and graphophone.

It was also during the first decade of the twentieth century that the two companies refined their marketing strategies to include intensive training and education, and sold not just the recorder itself but also a set of prescribed

In the background of this advertising photo, taken around 1910, a letter writer uses his cylinder "dictator," while cylinders are transcribed by a typist using a "transcriber." U.S. Department of the Interior, National Park Service, Edison National Historic Site, West Orange, New Jersey.

practices for using it. Office dictation systems aided in the process of management by providing managers with tools such as ready-made instructions defining women's work. More than just operating instructions, these prescribed practices would, the manufacturers argued, maximize efficiency in the office, although from today's perspective it is also clear that if companies followed those practices they would increase profits for National Phonograph and Columbia.

The marketing strategies associated with the Dictaphone and Ediphone were based on observations of the operations of large, scientifically managed firms. Typically, two types of employees (both, incidentally, nearly always male) generated correspondence. These included a relatively small group of upper and mid-level managers, plus in some types of companies a much larger group of lower level clerks who dealt with customers through correspondence. Examples of the latter included insurance agents or salesmen who at that time conducted most of their business through personal visits or written communication.

As it happened, few male workers found the use of the dictation machine pleasant. Many complained that they felt uncomfortable talking to the machine. However, lower level employees were more likely to use dictation equipment than their superiors simply because they were ordered to, or they were denied human stenographers. Upper level employees could often negotiate their way out of compulsory use of the machines, and the top managers had access to personal secretaries who were on hand to take dictation. The status conferred by having a personal secretary was one of the perquisites of upper management, and this was difficult to dislodge with a machine. Yet some men at all levels were drawn to the machine, and used it willingly (even enthusiastically) to record correspondence.

However, most companies went without dictation equipment or made its use optional. Middle-level managers and lower level letter writers would call on a stenographer from a secretarial pool to take dictation. Typists then typed up or "transcribed" the resulting letters using the new typewriter technology. Dictation machine promoters went after these companies aggressively, promising an increase in efficiency resulting from the elimination of all hand stenography. A letter that was recorded could be taken (in a special cart designed for that purpose and sold at extra cost with the recorder) directly to the typing pool without passing through the hands of a secretary or stenographer. There would be no mistakes due to poor note taking, no waiting for a stenographer from the pool to arrive, and no delays in transferring shorthand notes to the typing pool. While that was the argument, few listened.

A MATURING TECHNOLOGY

While the dictation machine had improved markedly since the 1880s, it retained one key element from the past—the cylinder. The disc revolution that swept the entertainment industry after 1900 left the business recorder market untouched. Almost incredibly, the wax cylinder survived the triumph of the disc and was going strong even after World War II. The last new cylinder Dictaphone model appeared in 1950, and cylinders and related supplies were readily available into the 1960s. Part of the reason for this conservatism in design was related to the way the dictation machine had evolved into a piece of equipment suited only for office use. After 1905, both the Ediphone and Dictaphone lines came to incorporate machines optimized to the various tasks undertaken in dictation, rather than being like the all purpose recorder-players that had come before. Now there were to be "dictator" models for men to use to make letters. These machines could record and reproduce, and were typically started and stopped with a control built into a handheld speaking tube (electric microphones were not available until the 1930s). Speaking tubes and microphones were designed so that they worked best when the user spoke somewhat quietly, with the mouth very near the microphone. The recorder was suitable only for voice recordings (as the occasional use of a dictating machine to record music demonstrates), and the machine could tolerate flawed grooves and rough handling unlike the sensitive recording machines used in music studios.

The secretarial models, called transcribers, were playback-only devices fitted with foot controls to free the hands for typing. Rather than having horns for listening, manufacturers returned to the old listening tube idea, not because the recordings were too weak to be heard through a horn, but because it was sometimes difficult to hear the words distinctly over the clack of the typewriter without increasing the volume to distracting levels.

Increasing mechanical specialization would, it was hoped, allow manufacturers to fully exploit what was seen as a huge potential market. While some companies, such as Sears, Roebuck and Company and several of the larger New York insurance firms, made dictation machine use a mandatory part of their office practice for clerks, the market was still quite limited by 1910 relative to the enormous number of business enterprises in the United States and around the world. However, profits per machine were high because the equipment was expensive, with a recorder being comparable in price to a typewriter, and because a typical sale included not only a recorder and a transcriber but also the furniture on which they sat, a cylinder shaver, carts for the cylinders, preprinted labels, special wax pencils for marking cylinders, and other accessories.

Despite the fact that Edison sold only about 5,000 units in 1910, it was the best year yet for the company, and salesmen were proud of their accomplishments. Sales would continue to grow, with total Dictaphone and Ediphone sales surpassing 22,000 units by 1923. Significantly, some of the rising sales in the late 1920s were attributable to purchases by the federal government, which eventually embraced dictation technology and made it a standard feature of numerous agency offices. The government (including the military and the Veterans Administration hospitals) would become one of the most reliable customers for dictation equipment in later years. But in a larger sense, dictation machines remained a relatively insignificant product compared to either the entertainment phonograph or many other types of office equipment. The typewriter, for example, in most years sold in numbers in excess of ten times the sales of dictation machines.

Given the limited nature of the market, it is not surprising that the Dictaphone and Ediphone attracted virtually no competitors. While scores of companies made entertainment phonographs over the years, almost none attempted to break into office dictation until after about 1950. One of the few exceptions to that trend was significant not because of its commercial success (for it had little), but because of its remarkable new technology. This firm was called the American Telegraphone Company.

THE INVENTION OF MAGNETIC SOUND RECORDING

American Telegraphone (the word is pronounced with the emphasis on the second syllable) was the U.S. licensee of a new kind of recorder, invented by a virtually unknown Danish engineer named Valdemar Poulsen. Born in 1869, Poulsen was a self-taught electrical engineer who had worked for some years for the Danish telephone company. Like many young men of his generation, he took up inventing as a hobby in his off-hours. Some time in 1898, he discovered that he could record sound on a steel wire without creating any groove or indentation. The process used a microphone (at that time still called a telephone transmitter), which generated a wavering electrical current, called the signal, corresponding to the pattern of the sounds spoken into it. From the microphone, that signal traveled through a wire to a small electromagnet (that is, a piece of iron wrapped in a coil of wire, in which the coil and iron core become magnetized when a current flows through the coil). The electromagnet would then radiate a magnetic field that fluctuated rapidly, in synchronization with the incoming electrical telephone signal. Any piece of iron or steel that was held near the electromagnet would itself become magnetized, so Poulsen reasoned that by rapidly passing a long steel wire near the electromagnet, the wire would be magnetized along its length and would retain a record of the fluctuations. One could think of the process as being analogous to the way a motion picture film retains images of a moving scene along its length, except that of course the magnetic "image" was invisible. As in the phonograph, reproduction of the sound involved reversing the recording process. As the length of wire was pulled across the face of the electromagnet, a fluctuating flow of electricity would be generated in the electromagnet's coil due to the principle of induction. That flow would actuate a telephone receiver (substituted for the telephone transmitter in the circuit) to reproduce the sound. Elements of

A Poulsen Telegraphone made around 1910 and intended for use as a dictating machine. By permission of University Archives, Paul V. Galvin Library, Illinois Institute of Technology, Chicago.

the telegraphone or the magnetic recording process had been anticipated by Edison, Charles S. Tainter, and others, but never had this type of sound recording been reduced to practice and demonstrated to the public.

There is some evidence that the first magnetic recording device may have been conceived and possibly constructed by the prominent American mechanical engineer Oberlin Smith in the 1870s, and while Smith in 1888 published his ideas in a well-known technical magazine, he did not patent or demonstrate a working model of his recorder.[1] Valdemar Poulsen may have read about Smith's device, or he may have reinvented it. In either case, his first recording was apparently made when he strung a long steel wire across an open field and held the microphone/electromagnet circuit in his hand, pressing it against the wire and running along its length while he shouted into the microphone. It must have looked ridiculous, but it worked. He filed for patents covering the magnetic recording process between 1898 and 1900 in Denmark and the United States. A few newspapers and magazines ran stories on the remarkable new machine as early as 1899, but it was Poulsen's demonstrations at the 1900 Paris Exposition that propelled him to near-stardom in the engineering and scientific communities.

Poulsen's device violated the principles of physics as they were commonly understood in the late nineteenth century. Poulsen explained that the recording on a wire consisted of varying regions of magnetization, which differed from one another in strength and the orientation of their north and south poles. But according to the theory of magnetism accepted at that time, a magnetized piece of iron or steel always had a uniform level of magnetization. "Spot" magnetization was not possible, because a magnetized spot would spread until the entire body was uniformly magnetized. By this same logic, one could not magnetize a piece of metal in two different places with the magnetic poles oriented in different directions. But Poulsen's recording process depended on the creation, in effect, of numerous magnets oriented in multiple directions along the length of a wire. In fact, spot magnetization is possible in magnetically "hard" steels, in which a strong magnetic force is required to magnetize them in the first place. The atoms in these steels, once magnetized, tend to stay that way even in the presence of another nearby magnetic force.

Between the time of the 1900 demonstrations in Paris and about 1915, Poulsen designed several different models of the telegraphone. The first and simplest looked a great deal like an oversized cylinder phonograph. A large metal cylinder, scribed with a spiral groove, was wrapped with a length of recording wire. A combination recording-reproducing electromagnet tracked the wire as the cylinder was rotated. This was the simplest type of telegraphone, but the least useful because it could only record a few seconds of sound. He later demonstrated a machine that used a thin steel tape rather than a wire. The weight of the tape made this machine difficult to keep ad-

1. The assertion that Smith invented the device in the 1870s is substantiated only by evidence from 1911, according to Cox and Malim, *Ferracute* (1985). See also Oberlin Smith (1888).

justed, since starting and stopping the reels put a great strain on the mechanism. A third basic type of telegraphone using reels of thin wire seemed the most promising. Fitted with lengths of wire to allow up to 30 minutes of total recording time, this was the machine best suited for office dictation, the application that Poulsen suggested for it. In a demonstration conducted in the United States in 1900 or 1901, the telegraphone proved to be an acceptable substitute for the business phonograph, and in fact the clarity of the recordings was said to be significantly better than that of the phonograph. Further, it could be attached to the telephone, making remote dictation possible and also enabling the recording of telephone conversations for records keeping or other purposes. In this respect, the telegraphone could not be equaled by the phonograph, which could not yet record the telephone reliably.

It was several years before American investors could be gathered together to form a company to license the right to manufacture and sell the telegraphone in the United States. The American Telegraphone Company was finally established in 1903 amid great publicity in the newspapers and technical magazines. By 1905, a significant advertising campaign was attracting millions of dollars in stock purchases, extracted from thousands of individuals across the country who thought they were being given the chance to get in on the ground floor of "the next phonograph." As evidence of the success of this offering, original American Telegraphone Company stock certificates were still readily available from antiques dealers in the early 2000s. At a time when scientific office management was taking hold and dictation machines were becoming more familiar, the advanced technology of the telegraphone appealed to many investors.

However, the management of the American Telegraphone Company proved unable to make working telegraphones at its original Wheeling, West Virginia, factory. A reorganization in 1908 resulted in the closure of the Wheeling factory and its reestablishment in Springfield, Massachusetts, but there were still no telegraphones produced before 1912.

Under a new regime, American Telegraphone produced perhaps a few hundred machines between 1912 and about 1918. The majority were sold to curious experimenters and competing companies, such as the American Telephone and Telegraph Company (AT&T), for evaluation in their laboratories. The only documented commercial installation was at the DuPont company, where twenty telegraphones were used in 1912 or 1913. These machines were installed in a room that served as a dictation center, connected to the desktops of letter writers by special telephone circuits. The machines seemed to have worked well for a few years, although they required considerable maintenance. By 1917 they were worn out, and when the company purchased replacements the second installation proved to be less reliable. The experiment was terminated in 1919, and after 1920 American Telegraphone was effectively out of business.

DIVERGING STREAMS OF INNOVATION

By 1920, the needs of the dictation market had led to a form of sound recording that was quite different than the technology being sold in the entertainment market. Entertainment was becoming the realm of the disc phonograph and the playback of records in the home. Using improved versions of Emile Berliner's disc gramophone, manufacturers and record makers had enticed Americans and Europeans to purchase inexpensive record players and discs. Sales of home *recorders* as opposed to players sank to a very low level and remained there for decades. Meanwhile, the cylinder phonograph had revived and become established as the standard technology for office dictation. Two American companies developed an entire system of phonograph-based correspondence around it. To a large extent, the technology of office dictation remained stable for the next several decades, until a new generation of magnetic recording devices spurred a host of innovations. But this would not be the direct result of the telegraphone, a technical milestone but commercial flop. The commercial successes of office dictation, though limited, were stimulating inventors to think of new ideas that would, in the long run, become important.

6

The Heyday of the Phonograph

THE SCENE IN 1910

In the period before 1910, the phonograph, graphophone, and gramophone all became established, improved in quality, and spawned a number of imitators. The phonograph and graphophone, both originally cylinder recorder-players, saw design changes that made them nearly indistinguishable, except when inventors or manufacturers would briefly test the market with long-playing cylinders or other innovations. Although cylinder sales grew between 1900 and 1910, the number of cylinders (and cylinder players) sold was soon overshadowed by the much faster growth in the sale of gramophone disc players. The cylinder would eventually disappear entirely.

THE RECORD INDUSTRY'S FIRST PEAK

The years from 1910 to about 1925 saw the "talking machine" (the term that increasingly came to refer to the disc gramophone) reach a peak in sales that would not be surpassed for many years. Responding to falling prices for players and discs, Americans and Europeans made this technology a part of their daily lives. The impact of recording on both popular culture and music itself was quite striking.

Filling cylinder molds at the Edison factory at West Orange, New Jersey, 1914. U.S. Department of the Interior, National Park Service, Edison National Historic Site, West Orange, New Jersey.

One of the most important outcomes of the introduction of sound recording—not only to the public but also to the recording industry—is traced to the series of recordings made by Enrico Caruso in 1902. These recordings have attained something of a mythical status in the industry and among record aficionados, because of the boost they gave to the talking machine's reputation as a serious musical instrument. Previously the device had been seen as a second-rate way to appreciate music, far inferior to the live performance. Some critics, notably the composer John Philip Sousa, denounced the phonograph as a threat to good taste, because it would in his view lead lazier people to rely on its inferior performance rather than troubling to attend an actual performance.

Sound recording companies would battle this negative image for many more years, but the Caruso recordings were the first step toward "legitimization." One milestone in this transition was the introduction by Victor of its famous Red Seal records, for which the company actually charged a premium, countering a downward price trend in the industry. Whereas an ordinary record cost between 25 and 50 cents, Red Seal records could cost $2.00. Starting from the Caruso recordings, record companies were able to attract more and more top opera and stage stars to make recordings. Today it is difficult to appreciate the mood of those times, but in that day some artists believed that their stage careers could actually be damaged if the public heard them on the talking machine. The reason: the very poor sound quality of the early versions of the technology. Recordings made in the decade after 1900 still sound quite primitive to modern listeners, but they had improved greatly in a few short years.

After about 1900, U.S. record companies began expanding their business in Europe and elsewhere. Edison and Victor were already well established in England, France, and Germany, while their licensees were making rapid progress in Russia, Italy, and later South America and Asia. Importers were also developing markets for talking machines in India, Australia, and elsewhere. However, the boom in phonographs in the early 1900s was most dramatic in the United States. It was particularly evident during the relatively brief period during and immediately after World War I. The United States entered the war late and the country suffered no physical damage during the war, unlike the European combatants. While American soldiers suffered greatly in the trenches overseas, at home the war created economic abundance as the United States manufactured and sold massive quantities of armaments. The result was a general economic boom and an upturn in consumer spending. Among the products that consumers purchased were talking machines and records. The 1920s were "roaring," and it was the talking machine that provided that sound.

MAKING RECORDS IN THE 1920S

The two essential tasks involved in making records—capturing the recording and mass producing the records—evolved at uneven rates between 1890 and the end of the 1920s. The first, capturing the recording, would at first lag and then rapidly move forward in the mid-1920s. The making of copies of records for sale to the public had undergone intense development earlier, in the 1890s, but after 1915 seemed to stagnate until about the 1940s.

A quick review of the history of places where recordings were made reveals that in the early days of recording, there was really no such thing as a recording studio. Any relatively quiet room could be used, or in some cases recordings could be made outdoors. Remember that recording took place as early as the 1850s, before the advent of sound *reproducing*. The early recording devices like the phonautograph were scientific instruments, so recordings were made in the laboratory or in lecture halls before an audience.

Edison's first phonograph recordings were made in his laboratory in Menlo Park, New Jersey. After copies of the recorder had been made, Edison and his agents demonstrated it across the United States and in Europe, again usually in lecture halls. The first commercial application of the phonograph was the ill-fated talking doll first made in 1889, but little is known about who made its recordings or how they were made.

It was Edison's company that probably made the first entertainment cylinders for distribution to licensed phonograph distribution companies. These were also apparently made in the expansive West Orange laboratory complex where Edison moved in the 1880s. After a while, Edison set up a permanent recording room in the laboratory. Similarly, the first recording studio set up by Berliner probably consisted of merely a room designated for that purpose. Only after recording became a more regular event did studios designed especially for recording emerge. The recording phonograph, graphophone, or gramophone was sometimes located just outside that room, with the recording horn poking through a hole in the wall.

The early recording technology was so unreliable that it required much practice and often several "takes" to get a good recording. One approach was to have the artist or artists perform while several recorders were operating, and hopefully some of the recordings would sound good. All sorts of things could interfere with the making of a good recording. The hardness of the waxy recording compound could change with atmospheric temperature or humidity, or it could contain impurities or bubbles that could cause the stylus to jump. Early recording studios were often heated to uncomfortable temperatures so that the wax blanks would remain soft. During recording, things had to remain quiet and relatively still. If the recordist (as recording technicians were called) or performers bumped into the machine, it might instantly ruin the whole recording. With disc recordings, where the movement of the recording stylus was side to side rather than up and down, shouting too loudly into the horn could make the stylus break into an adjacent groove. Alternately, if the volume was too low, the resulting recording would be weak or inaudible. Skilled recordists were highly prized employees who were barraged with offers from competing companies and who changed corporate alliances frequently.

Performers faced serious obstacles, too. Some types of instruments simply would not record well. Others, such as drums, were too loud and

Harry Anthony making a record for Edison, circa 1907–1910. Early studios were small and cramped because of the need to crowd close to the recording horn. U.S. Department of the Interior, National Park Service, Edison National Historic Site, West Orange, New Jersey.

had to be muffled. Singers accustomed to working onstage had to learn to modulate their voices differently for the phonograph, and had to remember not to gesticulate too much, lest they knock the recording horn. Groups of musicians had to be crowded near the horn to be heard, limiting the size of "orchestras" to just a few players and making it difficult to perform tasks such as turning the pages of sheet music without causing a disaster. However, the early recording studios were less demanding in other ways. The noises of breathing, paper shuffling, light footsteps, and whispering were usually too faint to be captured. The whisper-quiet atmosphere of the modern studio was unnecessary in the days of acoustic recording.

From today's perspective, the toughest part of recording was probably the fact that a song or performance had to be "perfect" the first time. It was not possible to edit the recording in any way. Nor was it possible to record part of a recording first and "mix" in a second part later. Musicians, particularly those working with bands, always came to the studio ready to perform flawlessly, and usually practiced their songs many times onstage before recording it. Of course, creating the "perfect" performance was an elusive goal. In one famous story, Louis Armstrong dropped his sheet music in the middle of a song and had to improvise the rest of his lyrics on the spot. He did it so well that the recording was released and his impromptu "scat" style inspired others to imitate him, spawning a new form of music. What was really a mistake created a "perfect" performance. More mundane mistakes such as missed notes could result in a ruined recording, but if they were very minor they would simply be allowed to pass. Since there were nearly always flaws, at the end of the session it was the recordists' responsibility to listen to the recording and decide if it was good enough.

"Acoustic" recording technology was gradually improving in terms of sound quality. In 1913, Edison finally introduced a disc record, and late in 1915 his company began a combination demonstration and marketing campaign called the Tone Tests. These public tests, held in music halls, challenged the audience to detect whether a performance was live or a recording. Both the artist and the recording of the artist on an Edison Diamond Disc were played, then both were hidden behind a curtain. Sometimes the audience could tell and sometimes it could not, and Edison's team cheated a little by carefully selecting artists who could mimic the sound of their recordings. Despite these faults, the Diamond Disc was an improvement, and the Tone Tests demonstrated that the acoustic recording process had come a long way from the nearly unintelligible tinfoil phonograph.

RACE MUSIC

One of the most important cultural influences of the talking machine was that it captured not just the music of the day, but also the underlying attitudes of the societies that created it. The story of ethnic or "race" records has, like the Caruso recordings, achieved a mythical status in the eyes of collectors and music historians, although for very different reasons. Race records, as they were then known, were recordings intended to appeal to the numerous working-class ethnic groups in the United States. With the rising incomes of the post-1900 period, more working-class people were able to afford talking machines. According to historian Andre Millard, since the large record companies had a tight hold on the market for recordings with a mainstream appeal, smaller record companies typically sought to specialize in particular niches. Many of them turned to selling recordings intended for America's minority (but still sizable) populations of

The cover of this Columbia Phonograph Company catalog from around 1925 demonstrates the way that African-American music was marketed to mainstream audiences. Library of Congress, American Memory Collections, Washington, D.C.

Eastern European and Jewish immigrants. Another approach was to try to market records to America's rural population, which was in these years about half the country. Rural Americans, who often had different musical tastes than city folk, had been largely ignored by the record manufacturers (with the exception of Edison's company) in early years in favor of the more concentrated urban markets.

Race and ethnic records began to appear in greater numbers after 1900, but they were nearly always performed by white musicians. In some cases, they were presented to the public as authentic examples of ethnic culture, but often they were merely caricatures of "hayseed" or "coon" music aimed at amusing a white, mainstream audience. While some of these recordings were marketed to ethnic audiences, such as immigrants, to remind them of home, others were lampoons of one ethnic group for the amusement of another. For example, a song that ridiculed the Irish might be intended for the Jewish market. "Coon" (black) music was intended to amuse whites. In the context of the times, this kind of humor was common, and would not have been seen as being quite as malicious as it seems today. It was part of a long tradition in American stage and musical entertainment, where the culture of one ethnic group was parodied by another. However, it did tend to reinforce damaging racial and ethnic stereotypes.

While most of these forms of music have virtually disappeared, "race" records of African-American music persisted, although in a different form. Few African Americans actually made recordings before 1920, yet their music began to move beyond mere parodies. Little authentic "black" music was recorded until about 1920, but forms of music associated with African Americans, especially jazz, were becoming more popular.

Jazz emerged in the South around the turn of the twentieth century and migrated northward before World War I, following the migration of African Americans to northern industrial cities. The first known jazz recording was made by the all-white Original Dixieland Jass Band in 1917. About three years later, black singer Mamie Smith was recorded by the small record company Okeh. These and other recordings are credited with starting a fashion for jazz and blues (neither term was well defined) in the United States. Andre Millard has noted how the early jazz recordings such as those by the Original Dixieland Jass Band or the later New Orleans Rhythm Kings were a far cry from the music that had been performed live in the South in previous years. It had to be reshaped for the phonograph, so that songs ended within the time span allowed by discs and cylinders. Also, some instruments had to be downplayed or removed because they did not record well on the phonograph. Perhaps more importantly, nearly all the early jazz captured on records was "cleaned up" for white audiences. Live jazz was improvised and disorderly. Its lyrics and themes often had sexual overtones. In revising jazz and blues for records, the music was made to conform to standards set by the record companies, which assumed (probably correctly) that the white audience would not want to buy such raucous, amateurish, and sometimes offensive music. What jazz retained was the flavor of the original, along with its emphasis on percussion instruments and faster dance beats.

The blues music fad set off in a few cities by Mamie Smith in 1920 was also taken up by mainstream record companies, presented to white audiences, and in the process radically altered. While it was still difficult to define blues because it ranged considerably in form and style, it most likely was originally characterized by its lyrical content, which portrayed the somber African-American situation. It had been invented by southern African Americans as a form of live, improvised entertainment some time before 1900. However, the music had to be tidied up for white audiences and, in some cases, shortened to fit on records. Following on the heels of Mamie Smith, an even bigger African-American blues star was Bessie Smith, who recorded hundreds of songs during the 1920s. Other jazz and blues artists followed, particularly New Orleans–raised Louis Armstrong, whose post-1925 jazz recordings were widely imitated. The number of jazz and blues performers and bands exploded in the early 1920s, and the music spread to other parts of the world, especially Europe.

The growing popularity of jazz and blues was quite a phenomenon. It appeared at a time when the world economy was, in general, booming. In the United States, the "white" versions of this music came to symbolize an age that valued wealth and youth, and its slightly subversive, "outsider" nature captured the spirit of the decade's relaxed morality—at least for urban whites. In fact, the 1920s became known as the "Jazz Age." Not all of this was due to records, however. In major cities, especially New York and Chicago, many people first experienced the music not through records but through live performances at clubs and "speakeasies." Black artists and bands, who were rarely called into the recording studios, made their reputations through live performances rather than recordings.

Many music historians look back to the 1920s as a crucial turning point, and treat its jazz and blues recordings like sacred relics. Many of the celebrated artifacts of authentic African-American music date to this period. However, in worshiping these artifacts, music historians probably exaggerate the extent of their impact at the time. Jazz and blues recorded by African-American artists were still somewhat marginal phenomena, limited in appeal to a relatively small segment of listeners. The most popular record during the 1920s was recorded by Paul Whiteman, who was indeed a white man, and his music was a barely recognizable variation of the jazz recorded by blacks. Whiteman assembled a formally trained orchestra and performed rather restrained popular tunes. He called it "jazz," and it had a syncopated, danceable beat, but most of the similarity ended there. For better or worse, it was the Paul Whitemans of the world who were making the music heard by most people via the phonograph, radio, or motion pictures.

The cultural impact of authentic jazz and blues recordings of the 1920s was also muted by the sharp decline of the industry after 1920. Not merely jazz and blues but all forms of recorded sound (except that delivered via the new technology of motion pictures) were heard less and less. Just before this decline, between 1914 and 1918, the number of companies in the U.S. record industry had exploded from 18 to 166, and the value of their products had grown from just $27 million to $158 million. Yet the growth in competition, combined with other factors, led to a sharp decline in sales in the early 1920s. The year 1920 was the best year for Edison's record company, which achieved sales of $22 million. Edison watched as his sales dropped by more than 100,000 units to just over 30,000 phonographs between 1920 and 1921. Victor lost half its sales in that same period, and Columbia filed for bankruptcy. The rest of the decade was marked by gradually deteriorating conditions (Millard, *America on Record*, 1995, 72–74).

Part of the explanation for this was radio. Broadcasting first appeared in most major cities around the world in the early 1920s, and by 1921 there were an estimated 250,000 radio receivers in use in the United States alone. For the first few years, many stations broadcast music from records, which helped familiarize people with new recordings, but apparently cut into record sales. As individual stations were brought together to form networks, the new broadcasting companies centralized their studios and began to depend on "live" music by radio. This seemed to suit the record companies, but it also cut even more deeply into record sales. High-quality music was now available in most areas via radio for free.

MAKING RECORDINGS IN THE ERA OF RADIO

The coming of radio and its challenge to the talking machine industry encouraged the record companies to try new things. The "integrated" record and player manufacturers, Thomas A. Edison Company, Columbia, and Victor, struggled to maintain their positions of leadership by promoting technical innovations. The most innovative was Victor, which replaced the technique of acoustic recording and in the process redirected the industry along a new trajectory. This transformation was an outgrowth of the development of a new technology called electronics.

In brief, the story of "electronics," the technology that is almost ubiquitous today in computing, entertainment, communication, and other fields, began in the late nineteenth century with the experiments of Edison and the English scientist Ambrose Fleming. Edison discovered and Fleming analyzed how a modified lightbulb can conduct an electric current through the vacuum of its interior. The flow is always from the hot filament (a source of electrons as well as light) to a positively charged wire or electrode located inside the bulb. The exact mechanism for the "Edison effect" was not well understood, and the device remained a curiosity for some years. In 1906, American inventor Lee de Forest took this idea and added the idea of a control screen, literally a wire screen placed between the filament and the electrode to intercept the flow of electrons. By injecting a tiny voltage to the control screen, the de Forest bulb (or "tube" as it became known later) could in effect increase or decrease the large electric flow from the filament to the electrode, giving the device a sort of electrical "leverage." De Forest called his device the Audion, because he could use the weak electrical output of a telephone as the screen-control signal. The large current flow from the tube varied in direct proportion to the screen-control signal—it was, in other words, a high-power copy of the original, weak telephone signal. This powerful output could do things such as drive a loudspeaker that could be heard clear across a room. The circuit that de Forest designed using the Audion was thus the first practical electronic amplifier, and in 1913 he sold the right to use the invention to AT&T.

Unfortunately the Audion was unreliable, and de Forest fundamentally misunderstood how it worked, so AT&T had to improve the tube itself as well as the circuits it was used in. By 1915, the company had done this and used the Audion to power the first generation of telephone "repeaters," the amplifiers that made coast-to-coast telephone calls possible in the United States. Meanwhile, it sought to commercialize other uses of the Audion, such as its use as an amplifier for radios or phonographs. The phonograph companies began experimenting with electronic amplification and also with using the Audion amplifier to make recordings. Edison's ill-fated experiments to use the telephone to drive an electromagnetic recording stylus had proven that ordinary telephone signals were too weak, but the Audion amplifier offered a way to boost those signals.

Several others inventors went to work on this problem, but no independent researcher or record company could compete with AT&T's research facility. Through its subsidiary, Western Electric, AT&T introduced an electronic phonograph recorder for record studios, as well as a new, scientifically designed record player for consumers. The first company to offer the new technology under license was Victor in 1924, but others quickly followed. Victor called its product Orthophonic, which means "straight sound," referring to the uniform response of the recording system at all sound frequencies. Edison (and several others) independently introduced similar systems within a few years.

In these new "electrical recording" systems, microphones and amplifiers replaced the recording horn used previously. If necessary, the output of several microphones could be "mixed" together electronically to form a single signal. By electronically controlling the sensitivity level of each microphone, amplifiers made it possible for musicians to spread themselves out in the studio, freeing up their playing style. Amplifiers could boost weak sounds or cut loud ones, and amplification made the recording system more sensitive, so that larger groups of performers could be recorded without losing the details of the sound. Because the radio industry had leapt ahead of the record companies in the early 1920s, much of what was needed to create the new electrical recording studios had already been pioneered in radio. The process of capturing and mixing sounds for radio broadcasts was nearly identical to the techniques used in recording studios, and most radio studios would have recording facilities on-site by the 1930s.

The use of the electrical recording process made a big difference in what consumers heard on records. Interestingly, though, the home players introduced to play the new records looked almost the same as before, since they still relied on acoustic playback and often retained their windup motors. Despite that, the system developed by Western Electric reproduced a wider range of frequencies even with acoustic reproduction, because of the careful design of the player. Victor and others later introduced players that ran with electric motors (these had been pioneered some years earlier for use on acoustic talking machines, but since a large proportion of American households did not have electricity until the 1930s, they were not the top sellers).

More innovative was Victor's introduction later of electrically powered record players that plugged into a special jack on the back of a Victor radio. The phonograph then shared the radio's amplifying circuits and loudspeaker. The key technical innovation was an improved electrical "pickup," which replaced the acoustic diaphragm and horn of the phonograph. The pickup was a device—a transducer—that converted the tiny movement of the stylus into a wavering electrical signal. This could be done in various ways, such as by attaching one end of the stylus to a tiny magnet, which interacted inside the pickup with a tiny coil of wire, resulting in the generation of a small signal that was then electronically amplified. By the end of the 1930s, the plug-in record player attachment or the stand-alone phonograph with built-in amplifier would virtually replace the old acoustic phonograph. But in the meantime, the phonograph industry had virtually disappeared. Edison dropped out of the phonograph business entirely by 1929 (although his company remained a leader in dictation machines). Columbia was absorbed by the new Columbia Broadcasting System (CBS) in 1934, and its phonograph business had shriveled. Interestingly, Columbia and CBS were separate companies with common roots. Columbia Phonograph had once consisted of American and English branches. The American branch went bankrupt in 1923, well ahead of the others in the industry. A leader of the English branch reorganized the American organization as Columbia Phonograph Company, which in 1926 bought a record company called Okeh Records. Louis Sterling, the leader of the new Columbia, in 1928 participated in the creation of the Columbia Broadcasting System. In 1932, a radio manufacturer called Grigsby-Grunow bought Columbia Phonograph Company from its English parent firm. Two years later, Columbia Phonograph was again bankrupt, and it was then purchased by Columbia Broadcasting in 1938 and reorganized as the Columbia Phonograph Corporation. Victor suffered a similarly humiliating defeat at the hands of the radio interests. The introduction of the Orthophonic system was not able to save the company, which sank year by year until 1929, when it was bought by the Radio Corporation of America (RCA). Nevertheless, the Columbia and Victor names would survive because of the support of the radio industry.

THE RECORDING INDUSTRY IN HIBERNATION

In their new forms as the Columbia Phonograph Corporation and the RCA-Victor Corporation, the leading phonograph companies were sheltered from dissolution by the much stronger firms in the radio industry. While sales of records would pick up by the middle 1930s, the industry was in turmoil through the beginning of World War II. In terms of innovation in recording technology, the phonograph companies were no longer at the forefront. In their place were firms, some of them new, which used sound recording in new ways.

7

The Talkies

SOUND AND MOTION

Even as the phonograph was fading into the background in the 1920s, an important new application for sound recording technology was beginning to emerge. Motion pictures were already well established by 1920, and a number of inventors had tried to link the moving image with the phonograph. That merger would prove to be more difficult than anticipated, and large sums of money would be spent trying to make motion picture sound work before a viable system emerged.

THE TECHNICAL BACKGROUND OF MOTION PICTURES

Today we still use the term "motion picture" to describe the technology that we experience in theaters. Yet nearly every motion picture also carries a sound recording with it, making it much more than just a "picture." The persistence of the term is probably due to the fact that when motion pictures first appeared, they were in fact just that: moving images with no sound. The making of motion pictures on photographic film dates from the late nineteenth century, but it was anticipated by numerous entertainment technologies dating back to the fifteenth century. Perhaps the first was the so-called magic lantern, a device that appeared as early as the 1420s, although the inventor is disputed. The device consisted of an ordinary lantern covered by an opaque screen. Light passed through a cutout shape (often a devil or mythical creature) in the screen, and the pattern would be projected onto a wall.

By the early nineteenth century, magic lantern shows were a significant form of commercial entertainment in Europe and the United States. The more elaborate shows used expensive, animated slides that projected what appeared to be moving images or shapes. There were also "phantasmagoria" shows that projected demons or other images onto smoke released into the theater. These gave audiences the sense of seeing floating, ethereal beings, and were in some sense a form of three-dimensional motion picture. There were also by this time magic lanterns that used a series of images, shown in rapid succession, to simulate an image in motion, such as a moving human figure. These depended on the fact that an image striking the human eye will persist for a fraction of a second, giving the brain the impression of continuous motion even if the actual "motion" is quite irregular. This principle has been used in motion picture technology ever since.

As magic lantern entertainment was hitting a peak in popularity in the late nineteenth century, photography was also developing at a rapid pace. Photography is based on the reaction of silver nitrate or other chemicals to the presence of light, a phenomenon discovered by the ancients but exploited for making pictures much later. Photography also depended on the "camera obscura," developed no later than the early 1500s. This device usually employed a room or large box with a tiny hole in one wall. If the light outside was bright enough, a faint image of the scene outside would be projected onto the rear of the room through the hole.

In the late 1820s, several inventors developed ways to capture images on plates or sheets coated with various chemicals that changed color when exposed to light. Yet the exposure process for early photograph technologies could take hours, so others looked for ways to shorten this time. A partial solution was the daguerreotype, invented by the Frenchman Louis Daguerre some time after 1892. This used a glass plate, coated with a new light-sensitive compound that required only a half-hour exposure and that could be "fixed" (that is, the exposure process could be stopped) at any point by a simple chemical bath. The career of the daguerreotype was limited, however, because other inventors came up with more appealing solutions. One aim of these inventors was to be able to make multiple copies of an image from a single original, which the daguerreotype could not easily provide.

This led to the invention of the "negative," which not only allowed but also required that at least one copy be made from the original in order to view the image properly. English inventor William Henry Fox Talbot had developed this process in the 1830s, and by 1840 had a viable system.

His calotype, as it was called, was itself subjected to numerous improvements and also became the basis of a growing commercial portrait photography business during and after the 1840s. An important innovation was Frederick Archer's 1851 "collodion" process, which involved using a new type of chemical coating on a sheet of glass (which became the negative) and which reduced the necessary exposure time considerably. The collodion process required that the photographer carry more equipment, because the plates had to be coated and processed at the time the photograph was made. However, this did not slow the commercial expansion of the photography business.

PHOTOGRAPHY IN MOTION

Several years after the invention of the phonograph, Edison would begin to try to "do for the eye what the phonograph does for the ear," using photography to record moving images. While a photograph-based motion picture system had been attempted as early as 1879 and a practical motion picture camera became available in 1895, Edison would make important contributions toward a practical commercial system.

Perhaps he anticipated this as early as 1883, when he hired photographer William Dickson. By 1888, Edison had produced a preliminary patent specification, called a caveat, that described a motion picture system. Dickson and others, building on the work of the French Lumiere brothers, had by 1899 designed a new motion picture camera, which was to use rolls of film purchased from George Eastman's company in Rochester, New York. Edison's team demonstrated an experimental motion picture accompanied by a phonograph later that year in West Orange, New Jersey. Edison also incorporated many existing inventions into the system he used for projecting the image onto a distant screen, but as in the case of the camera, his key contributions were related to making a product that could be used for commercial purposes, and then actively promoting the new technology to the public.

Unfortunately, the link between the phonograph and the motion picture was cut shortly afterward. Edison's kinetoscope was both a camera and a device for viewing the films (they were not projected onto a wall) and was successfully demonstrated in New York City to the public in

Edison's kinetophone of 1913 linked a phonograph to a motion picture projector. U.S. Department of the Interior, National Park Service, Edison National Historic Site, West Orange, New Jersey.

1894, but his kinetophone sound motion picture system was not a success in the laboratory and the idea was temporarily put aside around 1894.

Others also went to work on synchronizing the phonograph to a motion picture. As motion pictures achieved greater and greater commercial success, more inventors entered the field trying to improve the technologies for making them. There were literally dozens of proposals for linking sounds to the moving image, all of them impractical at the time. The main failure of Edison's system was the fact that the projector (he began offering projectors after briefly relying on the "peep show" kinetoscope) and phonograph would get out of synchronization. In a later version of the kinetophone system, which projected the image on a screen, the phonograph had to located in front of the audience for it to be heard, but the projector was at the rear. A long belt linked the two, but it was far too difficult to keep them synchronized. Others proposed putting the recording directly on the film. There were dozens of patents, for example, on systems that carried the sound in a groove along the edge of the film. These worked in principle, but no inventor ever came up with a suitable way to duplicate such a recording; duplicating the original film for distribution to theaters was by the late 1890s an essential part of doing business.

Still other inventors sought an entirely new way to link pictures and sound that did not depend on the phonograph at all. In 1899, Valdemar Poulsen invented the first magnetic recording devices, capturing sound on a length of steel wire. Several inventors proposed synchronizing the Poulsen recorder to a motion picture device, or simply embedding a recording wire or strip of steel along the edge of the film to carry the sound. However, before the advent of electronic amplifiers, the output of a magnetic recording was very weak, and there was no way for a large audience to appreciate a film made this way.

French-born inventor Eugene Lauste, who had worked with Edison and Dickson, independently invented a way to record sound as an image on photographic film around 1906. Lauste used a sensitive mechanical device that moved or vibrated in response to sound waves. When this device was placed between a light source and a strip of film, the vibrations created variations in the light, which could be captured as a photograph. By using a very long strip of film and moving it rapidly past the recording unit, a continuous record of the sound was created along the length of the film. Sound could be reproduced by shining a light through the developed film so that it struck a selenium cell. The selenium cell was an early form of semiconductor device, similar to today's solar cells in that it produces a tiny electric current when light strikes it. The varying intensity of light from the sound film, when striking the cell, caused a varying electric current to be generated in the selenium. In the Lauste system, that current could then be used to reproduce the original sound in a telephone receiver (or what we could call a headphone today). The system, or something like it, would eventually supercede all others, but in the meantime the best compromise was still the phonograph, because like the Poulsen wire, the volume of the playback from the Lauste system was too faint to fill a room, and by this time motion pictures were already being exhibited to groups of people in theaters.

Throughout the period from about 1900 to the mid-1920s, phonograph-based motion picture sound systems were in commercial use throughout the United States and Europe. While improvements in their design lessened the synchronization problem, they still gave theater owners headaches. In the United States, where the motion picture industry achieved the greatest commercial success, the most prominent studios stayed away from sound technologies. In later years, as the industry moved from its original home in New

Jersey and New York City and created "Hollywood" in southern California, it temporarily left sound behind.

The lack of recorded sound accompaniment did not stop the Hollywood studios from creating a booming market for motion picture entertainment. In fact, when an improved sound system appeared in the 1920s, many studio executives were indifferent to it, as silent films had proven their entertainment value. One should note, however, that while these early films were silent, the theaters were not. Films were supplied with musical scores, which were then played in the theaters by small orchestras or piano players. There were even player pianos installed in some theaters. So movies and sound were already performed together, but the nature of that sound was about to change.

The new sound technology of the late 1920s was developed by a number of inventors, but most successfully by the Western Electric Company, then the research and manufacturing arm of American Telephone and Telegraph. Western Electric had recently introduced a new, electronic system for recording and reproducing records, which it had licensed to Victor and other phonograph or record companies. Edward B. Craft, an engineer at Western Electric, then led a project to adapt this technology for the motion pictures. An experimental version of the resulting Vitaphone was demonstrated in 1922, and again in 1924; the tests seemed to justify Western Electric's belief that a high-quality, disc-based theater sound system was now practical.

The mature system, as demonstrated in 1924, relied on several new Western Electric innovations. Recordings were made in a carefully constructed studio using improved types of microphones and the new technology of electronic amplification. This provided a record with a wider frequency range and lower distortion than before. The recordings were made on 16-inch-diameter discs run at 33⅓ rpm. The speed was chosen to simplify the design of a mechanism that kept the phonograph in near-perfect synchronization with the projector. At that speed, the 16-inch diameter allowed a recording time of about 10 minutes per side, which was the same duration as a standard reel of film. Normally, even in silent movies, feature films consisted of several reels of film shown in succession. While the large discs would never be offered for sale to the public, the 33⅓ speed was later adapted for the long-playing (LP) record in the late 1940s.

In addition to the recording system, Western Electric had devoted considerable attention to theater reproduction technologies. Theaters would be fitted with electronic amplifiers and large loudspeakers, rather than relying on the old acoustic phonograph. The result was that even very large theaters could potentially show sound films with good sound at sufficient volume for everyone in the audience to hear.

Western Electric's system was in many ways simply an improvement on earlier technologies, such as Edison's kinetophone or inventor Lee de Forest's electronically amplified motion picture sound system. It was also a conservative design, relying on tried and true "talking machine" discs instead of the newer technologies that recorded the sound directly on the film. It was not the first system to use discs for motion picture sound, but unlike earlier attempts, it worked well.

THE JAZZ SINGER

There was indifference or even opposition to the introduction of the Western Electric system. On the one hand, musicians rightly saw it as a threat to their livelihoods. Theater owners saw little need for such an expensive new addition when films were already quite profitable. Most importantly, motion picture producers had tried and rejected numerous sound systems over the years and were highly skeptical of this new one. Western Electric demonstrated its system to the one studio executive who seemed interested, Samuel Warner, in 1925. Warner, with help from Western Electric, set up a studio in Brooklyn, New York, and made the first feature film using the system, *Don Juan*, in 1926. While this film was well received by the public, its success was overshadowed the following year when Warner released *The Jazz Singer* starring Al Jolson. The movie brought in enormous box office receipts, and its success forced all the other major studios to adopt sound as soon as possible.

Like many of the early "talkies," *The Jazz Singer* had much less spoken dialog than a modern movie. The recordings for *The Jazz Singer* mimicked the old style of film (now retrospectively known as a "silent" film) in that it featured a musical score designed to emphasize the action on screen. But the film's producers accentuated the new technology by staging dramatic scenes in which the actor's voices could suddenly be heard over the music. The popularity of this new "talkie" was such that soon films were dominated by dialog, and music receded into the background though it remained very important for its ability to heighten the audience's emotions. There was a brief but tragic period when some silent film stars saw their careers ruined because they lacked good stage voices or had thick foreign accents. But many survived the transition, and the use of sound provided new opportunities for those who could sing as well as act.

It was not only the talking and the improved synchronization between action and sound that won the public's heart. The Western Electric system used cutting-edge electronic technologies and recording techniques, and provided noticeably better sound than ordinary recordings or the radio. Part of the key was the advanced technology, but Western Electric also demanded that theater owners install the systems in a prescribed way. Some theaters were simply unacceptable and did not receive the systems. In short, Western Electric forced theater owners to consider "acoustical engineering" issues, often for the first time. Another significant factor, especially for small-town theaters, was simply the fact that the slick production values of the early talkie sound recordings often outclassed the abilities of local musicians in providing quality music.

When *The Jazz Singer* was released, only about 100 U.S. theaters had the Vitaphone system installed, and it was so expensive that many theaters merely rented it once to show that film and then went back to ordinary silents. But between 1926 and 1929, theater chain owners spent millions of dollars leasing or purchasing sound equipment. A single installation cost between $5,000 and $20,000, and the high cost of transition to sound had the unintended effect of forcing many independent theaters out of business and encouraging greater consolidation in the industry. By about 1930, more than 4,000 U.S. theaters had sound equipment installed worth over $37 million (some of this included ongoing service contracts), and 95 percent of the films being produced in Hollywood were talkies. Western Electric's virtual monopoly did not last long, however, because inventors had by this time brought the so-called sound-on-film system to the point where it could successfully compete with disc-based systems ("Sound Picture" 1931).

The new technologies for sound on film all employed some variation of the older idea of recording sound as a visible "track" on motion picture film. The Fox studios, in response to the Warner sound films, obtained rights to the patents of a number of inventions, chiefly those of Theodore Case, who had developed optical sound-on-film systems. Fox would use Case's technology to produce its Movietone films, mainly for newsreels, by 1927. The young Radio Corporation of America, affiliated at the time with General Electric, was also readying a sound-on-film system called the Photophone. Important innovations had also emerged in Europe, such as improvements in the optical sound-on-film technologies introduced in 1918 by Josef Engl, Joseph Massole, and Hans Vogt in Germany, and the Tonfilm system of Valdemar Poulsen and Peder Pedersen in Denmark, which was announced in 1923. In 1930, at the height of popularity for the sound-on-disc system, there were approximately 200 competing film sound systems on the market throughout the world.

Yet industry leader Western Electric was not sitting idle. Even before it introduced the sound-on-disc system, researchers at the newly created Bell Telephone Laboratories (the successor to Western Electric's research department) were working on their own optical sound-on-film system. They introduced this system commercially around 1930, later distributing it through a new company called Electrical Research Products, Inc. (ERPI). It too was expensive, but it eliminated the minor technical problems associated with discs (even though it introduced a few of its own), such as breakage, skipping, and premature wear. For theater owners who had already installed the disc system, conversion to sound on film was expensive but not too painful. Existing projectors could be converted simply by installing a new, optical sound pickup "head." Much of the rest of the theater sound system could remain in place. RCA's competing Photophone system proved to be the major competitor, largely because RCA had established its own chain of theaters using the technology. Happily, the Photophone and Western Electric systems differed mainly in the way they recorded sound, not in the way they reproduced it. Any projector equipped with an optical sound pickup could run films with either type of soundtrack. By about 1936, many sound-on-disc systems in theaters had been supplemented by sound-on-film attachments; the triumph of sound on film was complete, yet it went nearly unnoticed by the public since it offered them almost nothing that was new. It was mainly important to the theater owners who owned and operated the equipment and to movie producers who developed recording on film into a new art in future years.

The transition from one type of theater sound system to another was less noticeable to the public than the dramatic downturn in the motion picture industry. The effects of the Great Depression that began in late 1929 were delayed by several years, but by about 1934 many of the major studios were bankrupt. Those remaining had to trim back their output and concentrate on making the types of films that brought the highest profits. The studios found one quite successful Depression product in the form of elaborate musical productions. These films, historians have argued, offered an escapist fantasy to moviegoers whose own lives were increasingly dreary.

The growing size of movie sets and the emphasis on music encouraged the continued improvement of recording and reproducing technologies. There were not many major breakthroughs, but there were continual refinements in studio design, microphones, electronic amplifiers, and loudspeakers. Also, the expanding use of recording technology in the movies provided opportunities for inventors. When sound recording had been limited to the phonograph, inventors who developed new ways to record or reproduce sound faced a difficult battle to convince the masses to make a switch. The inertia of what today would be called the "installed base" of consumer technology, be it the Victrolas of the 1920s or the CD players of the late twentieth century, usually militates against rapid changes of format.

Those who invented new forms of the sound recording machine at that time, such as Valdemar Poulsen, often expected to break into the nonconsumer sound recording market, but found that this was limited to office dictation. When the talkies emerged in the late 1920s, they offered the possibility of an entirely new and booming market for sound recording and reproducing technologies. With movies, consumers would not have to purchase a new piece of equipment, but would experience innovations in audio technology at their local theaters. Because so few studios controlled so much of the market, if they could be convinced to adopt a new technology it would surely succeed. This is exactly how the Vitaphone, Movietone, and other systems had attained success in the first place, and later inventors tried to duplicate the phenomenon.

The centralized, public nature of the theaters is also the reason why they tended to be the places where the public first experienced advances in audio technology during most of the 1930s and 1940s. Yet, to be sure, the number of viable inventions far exceeded the number of innovations that actually made it into the theaters. The story of the movie *Fantasia* is a revealing case study. The Walt Disney studios, which specialized in cartoons and children's entertainment, began making *Fantasia* in 1937. The film was to highlight its use of sound as much as its use of imagery. In fact, the first part of the film actually shows on the screen a slightly exaggerated version of an actual movie soundtrack.

It is, not coincidentally, the RCA type rather than the AT&T type of soundtrack, as RCA participated in the making of the film. The presentation of the animated soundtrack on screen was intended to emphasize the fact that the film featured something Disney called Fantasound. Yet when RCA engineers designed the Fantasound system, they were working with a technology that had been invented at the Bell Telephone Laboratories in the mid-1930s. Bell Labs engineers had demonstrated a recording of a full orchestra, recorded optically on motion picture film using multiple optical tracks rather than just one. The combination of these tracks, played back simultaneously through three separate sound systems, presented the listener with an aural illusion, a certain sense of spatial "depth" that is familiar to those who have listened to today's stereophonic recordings.

Building on the Bell Labs system, RCA engineers made extensive modifications and helped Disney improve studio recording techniques to enhance the effect of the multitrack recordings. The equipment used to exhibit these films in theaters cost approximately $100,000, and was so expensive that Disney representatives decided to make *Fantasia* a "road show," like the original *The Jazz Singer*, taking the equipment from town to town and premier to premier. The film did not draw large numbers to the theaters in

1940, but ironically after it was rereleased in 1942 with an ordinary soundtrack, it gained a following and continued to be revived periodically for many years. The lessons of the *Fantasia* case would not be taken to heart in the industry, and the public's lack of enthusiasm for innovative sound techniques would reappear repeatedly in future years. Theater owners were also reluctant or simply unable to afford such expensive upgrades to their sound systems, but these upgrades were regularly pushed on them by motion picture producers. When, for example, the Hollywood studios switched to magnetic soundtracks after World War II, many theater owners simply refused, and as a result motion pictures were released with optical soundtracks through the end of the twentieth century.

HOLLYWOOD, SOUND, AND THE AUDIENCE

There were significant but largely unseen implications of the merging of the motion picture and the sound recorder in the motion picture industry. The disc system introduced by Western Electric was obsolete almost as soon as it appeared, and was quickly replaced by sound on film. But the impact of the new optical recordings was more significant in the studios than the theaters. Optical sound recording provided significant advantages to moviemakers who were increasingly interested in editing and rearranging segments of film to produce a more elaborate product. Once optical sound was introduced, producers found that it was well suited to the economic climate of the 1930s, when musical productions and other big-budget films were produced in greater numbers. Further, the techniques and hardware developed by moviemakers for recording and reproducing sound would strongly influence recordings made outside the industry from the 1930s on.

8

Records and Radio in the United States

SOUND AND SPACE

The advent of radio broadcasting was a major turning point in engineering and social history. It marked the beginning of a new era in communication, linking ordinary people, who were spread across vast geographic areas, to the broadcasters, who became sources of information, style-setters, and opinion-shapers. Radio also hastened the emergence of the new field of electronics, which has in later years spawned diverse technologies from satellites to computers. As radio developed, it shaped and was shaped by changes in sound recording technology. Like the motion pictures, radio technology gradually merged with recording technology, and today they are inseparable.

ORIGINS OF RADIO

Inventors imagined a technology like radio long before true radio was actually possible. Even in the late nineteenth century, there were many attempts to use electricity for "wireless" communication, in analogy to the wired communication that had been possible via the telegraph since the early 1800s, and then after 1876 through the telephone. While there are several claimants to the title of the "inventor of radio," several of the experimenters depended on the "inductive coupling" of low-frequency waves or the conduction of electricity through the ground rather than through space. While it was possible to achieve wireless communication in these ways, messages could not be transmitted over long distances except in extraordinary circumstances. What inventor Heinrich Hertz discovered around 1890 was that a simple electrical spark blasted out not only light and noise but also invisible high-frequency radio waves. Such spark generators became the first radio transmitters. Radio waves could not be seen or felt, but they could cause small currents to flow in nearby metal objects. Inventors discovered ways to harness these flows, translating them into something that a person could hear, see, or feel, such as an audible click in a telephone receiver, and these devices became the first radio receivers.

Thus the simplest radio systems generated sparks and detected the resulting waves at a distance, using this simple system to transmit information. The Morse code used in telegraphy was ideally suited to become the language of radio, since it consisted entirely of short "dots" and long "dashes." A spark generator, equipped with a simple on-off switch, could be "keyed" on for a brief period to send a dot, or a slightly longer period to transmit a dash. The bursts of energy, if strong enough, could be detected thousands of miles away. Government agencies, the military, and private companies began to use wireless telegraphy around 1910 for things like ship-to-shore communication or long-distance transmission of the news. Since all of this was done in Morse code, there was hardly any interaction between radio and sound recording, but that would soon change.

Within a few years, engineers had developed circuits for radio that generated high-frequency waves continuously. This had some advantages for wireless telegraphy, and its inventors saw those advantages in terms of reaching greater distances. Yet some would discover that the continuous wave equipment also made it possible to transmit voice and music as well as the simple bursts used in Morse code. While telephone companies briefly experimented with "radio telephony," it was an expensive substitute for ordinary telephone communication over wires. Further, there was no privacy with a radio-telephone conversation, because anyone with a receiver could tune in and listen. Then it dawned on people to turn this lack of privacy to their advantage, and they began thinking in terms of radio "broadcasting" (a term taken from the ancient agricultural technique for spreading seed). Reginald Fessenden in late 1906 broadcast sound from a radio station on the coast of Massachusetts, and sailors at sea were able to hear him playing his violin on their continuous-wave, radio telegraph receivers. Lee de Forest, the inventor of a simple electronic amplifier in 1906, the following year also broadcast sound via radio waves. The sounds he chose included several phonograph records, played on an acoustic record player with a microphone placed in front of the horn. The relationship between sound recording and radio had begun.

COMMERCIAL BROADCASTING

However, radio broadcasting remained in limited use until World War I, when military research and development pushed radio technology forward. By the end of the war, vacuum tubes for transmitters and receivers had been made more reliable and less costly, and better radio circuits had been designed. Radio stations, which had been largely unregulated, now had to be licensed by the federal government in the United States, or were established as a government monopoly in other countries. International conventions assigned channels that prevented the stations of one country from interfering with another, and also had the effect of creating a worldwide standard for the way radio signals were transmitted and received. What was then called "standard broadcasting" is still in use today, though it is called amplitude modulation (AM) in the United States to differentiate it from frequency modulation (FM), which came later. Broadcasting in the United States followed a unique path in the world, partly because it was left in the hands of amateurs and private enterprise instead of being made a public service. The result was the creation of a large number of competing stations. By about 1920, stations were springing up all around the country, sponsored by individuals or businesses eager to use the new medium for advertising purposes.

Many of these stations were small enterprises with few resources, there was suddenly a huge demand for inexpensive sources of audio content, such as news, sports, and entertainment. The phonograph record was an obvious way to provide some of that entertainment. Records formed a large part of what most stations "aired" during the period immediately after World War I, from about 1919 to 1922. In the large eastern cities, electrical manufacturers such as Westinghouse, General Electric/RCA (then affiliated), and American Telephone and Telegraph were now setting up "chains" or networks of stations. These chains linked together stations using telephone lines, so that they could share content. The chain, or "network" as it was later called, appealed to advertisers, because ads broadcast on these chains reached a large audience. The establishment of chains or networks soon began to have an effect on the type of content heard via radio, and whether it was live or recorded. Almost from the beginning, record companies sought to have their recordings played on the radio, but also sought to control the ways that broadcasters used them. Victor, for example, reached an agreement with AT&T in 1925 that allowed AT&T's small network of radio stations to play Victor records on the air.

Meanwhile, the exploding number of broadcast stations in the United States led to new complaints of interference between stations, or the disruption of military radio communications. Beginning in 1922, the federal government took a more active role in determining radio content in direct or indirect ways. Regulations that forced amateur stations off the broadcast band, for example, eliminated the stations with the least ability to sustain high-quality live entertainment and favored richer commercial stations. For a time, the government also sought to discourage the broadcast of phonograph records by creating a new type of license for the "better" stations, where the station agreed to minimize its use of recordings. Although the federal regulations for broadcasting were significantly revised after 1925, this initial discouragement of the use of records had long-lasting results. It reinforced the ongoing trend toward the formation and consolidation of large networks of stations, some of which had a nearly national reach. The impetus for consolidation was largely economic; it was expensive and impractical for individual stations to organize or create enough "live" content to fill the broadcasting day. The networks had the economic resources to gather plenty of talented performers, and could then offer it to local stations in a way that allowed both network and affiliate to make money. Clearly, though, these economic arrangements encouraged live broadcasts and made them the norm in the industry.

LIVE BROADCASTING AND THE RECORDING INDUSTRY

From central studios in New York City, Chicago, and Los Angeles, live performers would be brought in, or portable equipment would pick up content from remote locations and send it to the studios for redistribution. The enormous cost of all this was defrayed by broadcasting to as large an audience as possible, and the promise of reaching such an audience in turn made the sponsorship of radio programs attractive to advertisers seeking national or regional markets. The heads of the networks saw little role for records in all this.

However, there were still friendly relationships between the record companies and the radio networks. Victor and Columbia, among others, held recording contracts with famous stars that required the broadcasters to seek permission from the record companies to put those stars on the air.

The broadcasters, eager to feature those stars on the radio, let the record companies sponsor programs and agreed to advertise the record company with which the stars were affiliated. So, for example, Victor began a series of broadcasts on AT&T's flagship station, WEAF in New York City, beginning in early 1925, and continued later that year with broadcasts on rival WJZ, part of the RCA chain. Victor recordists also made copies of the WJZ broadcasts, perhaps for archival purposes, but also pointing the way toward Victor's future role as the recording arm for the National Broadcasting Company (NBC), successor to the RCA radio chain. By the time RCA bought the ailing Victor Talking Machine Company in 1929, the ties between the radio manufacturers, phonograph manufacturers, and record companies were growing stronger, even though the actual playing of records on the radio was by now rare, at least on the national networks.

LIMITED USE OF RECORDINGS IN THE 1930S

Just a few years into the network area, there were criticisms of the expensive system of distributing live radio content from a central studio over long-distance telephone wires, to be broadcast on local stations. Not every station in the country could become a network affiliate or had an audience large enough to make network broadcasts profitable. The more successful independent stations sometimes sought to form their own networks, but were often unable to gain affiliates in the best markets. This in turn limited their success and the kinds of programming they could offer. There were also independent producers of shows who wanted to distribute their material outside of the network system. In 1928, for example, two Chicago radio personalities began syndicating, or selling to other stations, their popular comedy show under the name *Amos 'n' Andy* (a show that is also remembered for its insensitive portrayal of African Americans). The show was recorded on discs and then distributed to any radio station willing to buy it. This method was relatively inexpensive and incurred none of the wire transmission fees needed to link studio to station (these had to be leased from AT&T, which had a monopoly on long-distance telephone service).

The time was right for the distribution of full-length radio programs on disc. The technology to make these discs had recently become available in the motion picture industry. Even before the changeover from sound-on-disc to sound-on-film systems in motion picture production and exhibition, Western Electric was looking for new markets for its 33⅓ rpm disc recording equipment, so it targeted radio stations and potential program producers such as recording studios. The long playing time of these discs, which could be extended up to about fifteen minutes on a 16-inch-diameter disc, was ideally suited to radio.

However, the live networks were at this time consolidating their positions and retained their domination of the distribution of radio programming through the 1930s. Even *Amos 'n' Andy*, a pioneer of the syndicated program on disc, was brought over to NBC in later years and broadcast live. Program recordings on the radio were mostly relegated to certain niches. Ordinary phonograph records were similarly given second-class status. There were some popular programs such as the *Make Believe Ballroom* that played records, and local stations routinely played discs at off-hours. Only a few sponsors of nationally advertised products, such as the Chevrolet division of General Motors Corporation, were attracted to the disc because of its cost savings versus dealing with the national networks. More commonly, commercial sponsors who wanted to air short ads on the radio but not create entire recorded programs would distribute their advertisements to stations on discs. These recorded ads would then be played during locally originated programs. The most memorable examples of these were catchy "jingles," which were played frequently between songs or other material. Finally, there were companies, such as the NBC Transcription Service, that offered libraries of generic recordings. These libraries were purchased or leased by stations to serve as background for programs, or for "filler" when other material was unavailable. Nearly every station thus used some recorded programming, but as one study discovered in 1930, a typical network affiliate broadcast only 11 hours of recorded programming in a 128-hour broadcast week.

CONTINUED OPPOSITION

The U.S. federal government, responding to a shift over the course of the 1920s from unorganized, mostly amateur broadcasting to large-scale, network broadcasting, revised its radio regulations. It reaffirmed its opposition to the broadcast of ordinary phonograph records on the basis that there were too few channels available to allow "canned" music to be used as program content. The rule was not absolute, but the new Federal Radio Commission made it clear that the renewal of licenses might be denied to stations, especially urban stations, not providing what they considered high-quality material, with the implication that ordinary phonograph records were not high-quality programming. The commission left open the possibility of broadcasting the longer program record-

The recording room at NBC's studios in New York around 1939. Radio networks had a complex relationship with the phonograph before World War II. Few played recordings on the air. However, most had recording facilities in their studios and a few produced program recordings for sale. Edmund A. LaPorte collection, Archives Center, National Museum of American History, Behring Center, Smithsonian Institution, Washington, D.C.

ings, which were not generally available for sale to the public; this was a tacit endorsement of those who were already making and selling syndicated programs on discs.

By the 1930s, these recorded programs were usually produced only on the large, 16-inch-diameter discs. The discs, made using Western Electric technology, were not usually playable on home phonographs, because after 1931 they used a vertically cut, hill-and-dale groove rather than the side-to-side laterally cut groove made standard by Victor years earlier. The broadcasters began calling these special discs "transcription" recordings. Both transcription discs and regular phonograph recordings now had to be identified as such by an announcer each time they were played.

Broadcasters and record companies were unprepared for a wave of opposition to recordings of all kinds from organized labor beginning in 1929. The movement was led by James C. Petrillo, head of the American Federation of Musicians in Chicago. Petrillo called for a complete ban on the broadcast of any type of recording on radio, arguing that records took jobs away from musicians. Petrillo rose to national prominence in the early 1930s as the champion of a profession that had been negatively affected by the technologies of recording. Individual recording artists sometimes also voiced their opposition to the broadcast of recordings. Record companies responded to pressure from their artists by putting statements on record labels that read "not licensed for radio broadcast" beginning about 1931. Petrillo's efforts were also aimed at a revision of the copyright laws, which enabled the collection of royalties for the playing of records though the royalties were only paid to the composers or publishers, not the performers or record companies. Royalties for the broadcast of transcriptions were established in 1932, but this rule did not apply to the playing of ordinary discs. Some recording artists also banded together as early as 1935 to try to exact royalties from radio stations playing their records, but the organization was not successful. The popular bandleader Fred Waring formed a similar pressure group in 1935, aimed at prohibiting the broadcast of records. He made headway by winning test-case court injunctions against the broadcast of his recordings at a station in Philadelphia, setting off a wave of similar lawsuits.

RECORDING TECHNOLOGY IN A MATURE BROADCASTING INDUSTRY

In general, the opposition to the use of records on radio subsided during the later 1930s. NBC, one of the three major American radio networks, officially broadcast its first recording in early 1937. This was a transcription disc made by a reporter who happened to be covering the arrival of the doomed *Hindenburg* zeppelin at its destination in New Jersey. Despite ongoing reservations about recordings, this was such a dramatic news event that the network could not resist the temptation to broadcast the largely garbled and incoherent description of the crash. Local stations, even network affiliates, were by this time using transcription discs extensively, and in later years would return to ordinary phonograph records as well. It is important to recall, however, that the original opposition to records was based not only on the concerns of government regulators and artists but also on the economic interests of the broadcasters themselves. The use of live, network programs was a business strategy designed to keep the network broadcasters powerful and profitable at the expense of independents. Business conditions would change dramatically after World War II, and with new conditions would come new attitudes about playing records on the air. Before that occurred, World War II turned the technology of recording in a new direction.

9

The Crucial 1930s

THE WORST YEARS EVER

The phonograph industry had seen its best years in the early 1920s and would see its worst years in the early 1930s. The most startling indication of the industry's downward spiral was the withdrawal in 1929 of Thomas A. Edison, Inc., from the very industry that the "Old Man" had invented. Edison himself would die just a few years later, probably believing that the home phonograph was obsolete. Sales of records in the United States, which in 1929 had been about $75 million, fell to just $6 million by 1932. The statistics for other countries were just as bad. Radio broadcasting had been the likely cause in the late 1920s, but by the 1930s the leading problem was the Great Depression. This worldwide event crushed retail sales for all categories of consumer goods. The automobile industry, for example, which had enjoyed tremendous sales growth all through the 1920s, saw dozens of firms close up shop forever. Even radio manufacturing was hit hard, with sales of home receivers falling from over 4.4 million units in 1929 to about 2.3 million by 1932. The phonograph and record industries worldwide were forced to merge or consolidate to stay in business. In the United States, Victor sold out to RCA, while Columbia was absorbed by the Columbia Broadcasting System. In Great Britain, the Gramophone Company merged with its rival, the Columbia Graphophone Company.

The Deutsche Gramophon company, the German arm of the Berliner organization, had severed ties to its parent firms during World War I, but would remain independent only until 1941, when it too was absorbed by another company.

Record companies had to scale back their operations and concentrate on the few market segments in which records would still sell. One of those segments was classical music, where buyers were wealthier than the average consumer and often collected large personal libraries of music. Because they preferred the best quality recordings, they would purchase expensive records more readily than the average consumer, who was more price conscious. In bad times, record companies would come to depend on these classical music lovers.

"HIGHBROW" MUSIC AND THE TALKING MACHINE

It was the classical music market for which many of the technical innovations of the early and mid-1930s were intended. In 1927, for example, Victor introduced its first record changer, a machine well suited to the listening habits of classical music fans. Record companies believed that classical listeners would buy changers because of the nature of the music. Unlike popular music, classical compositions did not always conform to the rigid time limitations of cylinders or discs, and compositions could sometimes last an hour or more. Recording classical performances to a single disc was nearly always a compromise, requiring that pieces be shortened dramatically or that mere excerpts of a longer performance be released.

The only way to provide consumers with more or less full classical recordings was to chop them into 3- or 4-minute segments, record them on several discs, and sell the discs as a collection. These sets of discs would then be sold in cardboard albums, and this is, incidentally, the origin of the term "album" as it applies to recordings. Yet with a conventional phonograph, the listener had to go to the player every few minutes to change the record. The automatic record changer, on the other hand, allowed the listener to load a stack of several records onto a long spindle, which held them above the turntable. On pushing a button, an internal mechanism would move the tonearm off to the side, allow the first record to drop onto the turntable, and then place the stylus at the lead-in groove. At the end of the record, the mechanism lifted and moved the tonearm and allowed the next disc to fall. These complex machines came with high prices; one early example cost $600.

The second way the record companies catered to the classical audience was to improve the sound quality of recordings. "Better" is always subjective, but recording techniques had resulted in arguably better records between the 1890s and about 1930. The introduction of "electrical" recording in the 1920s made it easier to record a broader band of frequencies, particularly those above about 3,500 Hertz (Hz, or cycles per second). But with the added high frequencies often came increased high-frequency noise, caused by factors including the rough surface of the records then in use. Ironically, it was during the late 1920s, when record sales were in decline, that many innovations in recording came into use. Victor, for example, developed an electronic circuit that suppressed some of the extraneous noise, and began grinding its filler (usually limestone or shale, which comprised most of the record) more finely so it created less noise. Columbia also introduced a low-noise record using a central paper core, laminated on each side with a mixture of plastic and very fine powdered stone.

A more radical departure was the Victor Long-Playing Record introduced in late 1931. This was not the successful "LP," but a short-lived predecessor of the same name. Derived in part from the discs then in use in the motion picture and radio industries, the new records were made of a low-noise Vinylite plastic and came in 10- and 12-inch diameters. With a speed of just 33⅓ rpm, smaller grooves, and closer groove spacing, it was possible to play a single side for up to 30 minutes, although this demanded a new record player with a special pickup. The individual disks were priced much higher than a conventional 78 rpm single, but because more music could be packed onto each disc, the prices of complete albums were less than half the price of multidisc versions on conventional 78 rpm records. There was no escaping the high prices of the players, however, which started at $350 versus less than 10 percent of that for a standard model.

There were, unfortunately, major technical flaws in these players. The slowly rotating turntable mechanism transmitted more unwanted low-frequency noise, or "rumble," to the amplifiers, and the lower speed was more difficult to control accurately, leading to audible variations in speed. Further, poor sales in 1930 and 1931 forced Victor to raise the prices of albums to a level that was about the same as the 78 rpm versions. The recordings themselves, some of them dubbed from existing recordings on 78 rpm discs, were not well reviewed by the press, contributing to the format's demise. The product was discontinued in 1933. However, this failed experiment was a demonstration both of the record company's attempts to cater to the classical audience and of the engineering fascination for what would be called "high fidelity."

HIGH FIDELITY IN THE STUDIOS

In the late 1920s, a few manufacturers of radios or phonographs began using the phrase "high fidelity" to refer to the way their equipment sounded. It offered, they claimed, absolutely perfect reproduction of whatever was on a record, or whatever was being broadcast. The term began to catch on among a small group of engineers and home listening enthusiasts. The gradual improvement of the sound quality of recordings was nothing new; it had been described in terms of better "tone" since the late nineteenth century, but was taking a new direction in the era of electrical reproduction. The downturn in the record business and the economy meant that few consumers enjoyed the benefits of high-fidelity innovations, though engineers working in motion pictures, radio, and recording pushed forward with them anyway in the 1930s.

There were crucial developments occurring in the industrial laboratories of the world's large electrical manufacturing companies. One of the most important was the improvement of equipment for studio recording. While electrical recording had been implemented nearly everywhere in the record industry by 1930, it had barely made its mark before the general economic downturn sent business into a tailspin. At RCA's main studios in Camden, New Jersey, the recording studios for many years were located on an upper floor, so that the recording turntables could be driven by weights on long ropes rather than electric motors. At the time the machines were built, electric motors could not yet provide perfect speed control, but during the 1930s electrically powered recorders were nearly universal at radio and motion picture studios.

While record studios had fallen behind, changes were on the way. In 1931, two teams developed a new type of recording device based on the concept of the "moving coil." One was an inventor working in England, who patented an improved record cutter for 78 rpm discs. That inventor, Alan D. Blumlein, worked for the new Electrical and Musical Instruments Company (EMI), the result of the merger between the British branches of the Graphophone and Gramophone companies. The cutting head used the lightweight coils of an electromagnet to push and pull the cutting stylus rather than the heavier iron armature used with the electromagnetic cutters of the 1920s. Because with a lightweight coil there was less mass in motion, the action of the cutter was more precise, resulting in a groove that more accurately represented the waveform of the original signal. The device was first installed at the Abbey Road studio complex run by HMV records in 1931. At nearly the same time, a second team was working independently on moving-coil cutters at the Bell Telephone Laboratories.

Their device was adapted for making transcription recordings rather than consumer records, meaning that it cut a vertical hill-and-dale groove rather than the lateral, or side-to-side, groove used on consumer discs. The famous conductor Leopold Stokowski, who had a strong interest in audio technology, participated in making some early recordings using the new Western Electric record cutter, some of which were released many years later.

Both versions of moving-coil recording technology were landmarks, though they were not familiar to the public at the time. Bell Labs' system went into production in the form of Western Electric transcription recorders and, along with changes in the materials used for discs, contributed to the improving quality of transcriptions in the 1930s.

Interestingly, both the Bell Labs team and Blumlein followed up on the moving-coil cutter with innovative, multichannel recorders. In December 1931, Blumlein filed for a British patent on a new system of stereo recording, which he called "binaural." The system used two channels, each with a microphone and separate amplifier to drive a cutting head with dual styli, set 45 degrees apart so that each wall of the resulting V-shaped groove carried one of the two channels. Test records were cut in 1933 and 1934, mostly of people talking while they walked back and forth near the microphones. These would test the ability of the system to produce a recording that would allow the listener to follow such motion by visualizing it.

At Bell Laboratories in 1928 and 1932, engineers made their first experimental recordings using special two-channel moving-coil cutting heads. The head incised two parallel grooves into a wax disc, so that if the input from two separate channels were fed to the recorder, the resulting disc could be reproduced in two-channel stereophonic sound. Bell Labs made several stereo recordings in 1932 of Leopold Stokowski and the Philadelphia Philharmonic Orchestra. Of the two systems, the Bell Labs version was the less elegant solution, since Blumlein's recorder put both channels into a single groove.

Unfortunately, EMI put an end to the stereophonic recording experiments, and Blumlein moved on to other projects. When engineer Arthur Haddy at Decca Records and others began working toward new stereophonic recorders in the 1950s, they rediscovered Blumlein's work. Although the inventor had died in an airplane crash in the early 1940s, his contributions were important to the development of the stereo discs that appeared in 1958.

Many of the improvements in disc recording that actually saw commercial use in the 1930s applied to transcription discs. The original transcriptions used a wax-disc recording process similar to that of an ordinary phonograph record, but with a different recording mechanism, larger discs, and the lower 33⅓ rpm speed. Important improvements were introduced by Western Electric in 1931, which included a new type of low-noise wax master disc and the use of Vinylite for the final pressed copies. The recording mechanism was revised to make vertical recordings like the original Edison phonograph, replacing the side-to-side recording briefly used earlier. Playback was through a lightweight, electromagnetic pickup with a jewel stylus. The result of these changes was to allow a frequency response (usually meaning the range of sound frequencies that could be recorded) up to 10,000 Hz. The next year, in 1932, RCA also offered Vinylite transcription records under the tradename Vitrolac.

Beginning around 1931, there were also several systems introduced for use mainly in radio stations that used solid aluminum discs as the recording medium. Although aluminum seems like an unlikely choice for recording, it was possible to emboss a recording into the surface of an annealed aluminum disc using a powerful amplifier, an electromagnetic recording head, and a special stylus held to the surface with several pounds of weight.

ACETATE

The introduction of the acetate transcription disc in 1934 was a crucial innovation that revolutionized the economics of recording. While Bell Labs and others had recorded on acetate earlier, the major force promoting this new technology was the relatively obscure firm known as the Presto Recording Corporation of Brooklyn, New York. Presto entered the business in 1930, when it offered a simple kit that converted a home phonograph so that it could record onto aluminum discs. The recording head was moved across the disc by a feedscrew mechanism resembling a lathe. In 1932, the company turned to experiments with aluminum discs coated with a layer of plastic. Now the aluminum served as the rigid base for the plastic, which became the recording medium. Using any brand of recorder, the Presto discs had excellent sound quality. Users found that the quality of recordings was limited more by the available recording machinery than the disc itself, and Presto began to build a growing business supplying the blanks. This was the first technology that created high-quality recordings that could immediately be played (up to about five times, after which they began to be noisy). The process was dubbed "instantaneous" recording, because the discs could be played immediately afterward without any further processing.

Another reason for the acetate disc's success was related to its low cost. The economics of transcription recording were similar to the economics of making regular phonograph records. A wax master transcription disc could cost $100–$150. Once processed into a master stamper, the master could be used to make copies that cost about $1.25 to $2.00 each in quantity. Because of the expense, it was unusual for a studio or a radio station to make a transcription recording unless it was going to be duplicated in quantity for wide distribution. The acetate disc, by contrast, cost little to buy and could be used once or duplicated in quantity. This was an important difference.

Radio stations and small studios began to use these discs for a variety of purposes, none of which had been economically feasible before instantaneous recording. These included making audition recordings of the acts of potential performers, making original programs for later broadcast, and making archival recordings of broadcasts known as "air checks." The latter became the basis of a new business in the 1930s for companies making off-the-air recordings of local stations, which were then sent back to program sponsors as proof that the local affiliates had indeed broadcast the material that they were paid to air. Though competitors entered the market almost immediately with similar products, rapid growth was enough to accommodate them all. Presto bragged that it was selling half a million blank discs per year by 1937. While a lacquer-based coating would eventually replace true acetate, the name "acetate" remained in use in the recording industry through the end of the disc era in the late twentieth century.

REVIVING HOME RECORDING

Edison, Columbia, and others abandoned designs that allowed home recording long before 1930. Most cylinders and all discs were made of materials such as hard plastic that could not have been recorded upon even if the recording attachments were available. Thus in the early twentieth century, it was relatively rare for consumers to make a home recording.

However, a small market for consumer recorders did persist, and inventors were convinced that it could be enlarged. One early home recording system, manufactured by the Edison Bell Company in England (probably not affiliated with either inventor), used a special wax disc and an electromagnetic cutter attachment for the home phonograph. Recordings made on the disc could be played a few times before they wore out, and then the wax could be rejuvenated by wiping it with a solvent.

RCA, as part of its effort to improve flagging sales, introduced machines as part of its 1930 line of record players that would allow the output of an electronic amplifier to be connected to an optional disc recorder. The new recorder utilized a small, pregrooved, 6-inch disc made from a piece of cardboard with celluloid plastic laminated to each side. Later, solid plastic blanks were sold in 10- and 12-inch sizes. The recording attachment used an electromagnetically driven stylus to emboss the recording into the pregrooved disc. By the later 1930s, the boom in professional instantaneous recording equipment may have been spilling over into the home market in a small way. The Remco Babytone recorder, for example, was introduced in 1936 at a price of $125. The 1937 Allied mail-order catalog (Allied was the predecessor of the current Radio Shack) advertised three recorders from Universal Microphone priced at $92 to $375. Manufacturers other than RCA eventually offered inexpensive phonograph-recorders, often sold as part of console radio-phonograph combinations. The finest of these later machines was probably the Wilcox-Gay "Recordio," introduced in 1941. Like most of its competitors, the Wilcox-Gay cut records at 78 rpm and could only handle 10- or 12-inch records, meaning that recordings had to be short.

REVIVING RECORDS

From the low point in 1932, U.S. record manufacturers saw double-digit sales increases every year from 1934 to 1937. The next two years were even better, with sales increasing 100 percent in 1938 and then another 68 percent in 1939. To put things in perspective, 1939 sales of $44 million were still less than half of the 1921 high mark, but much better than the $6 million seen in 1932.

Some of the uptick in sales had been stimulated by ever-lower prices for record players, beginning in 1932 with the RCA "Duo Jr.," a $16.50 attachment that connected to an RCA radio to share its amplifier and loudspeaker. Inexpensive models came to be the top sellers in the 1930s, with the average price of a phonograph from RCA, still an industry leader, declining by about half between 1936 and 1940. Many record sales at this time came from radio stations. Despite the continuing aversion of the major American networks to playing records, individual stations sometimes used them. In major cities, there was a fad in the late 1930s for shows hosted by "disc jockeys" who played the top hit records. By the end of the decade, it was becoming clear that these shows helped stimulate consumers to buy the music they heard.

The elaborate mechanism of this jukebox (this one from the 1950s or 1960s) held up to several dozen records on a carousel and used a complex mechanical system to locate, select, and place the proper record onto the turntable. David Sarnoff Library, Princeton, New Jersey.

JUKEBOX TO THE RESCUE

The coin-operated phonograph had given birth to the record industry in the 1890s, but with Prohibition in the 1920s, many outlets for coin-op machines in the United States had disappeared. They had persisted, however, in the Deep South where they were (according to myth) still heard in seedy "jook joints." With the repeal of Prohibition in 1933, these machines, now generally known as jukeboxes, came roaring back.

One of the most important manufacturers to capitalize on this was the Rudolph Wurlitzer Company of New York. The company had been a maker of pianos and organs for some years before it became the manufacturer of an existing automatic phonograph design in 1934. The Wurlitzer "Simplex" automatic mechanism could play twenty-four standard 10-inch discs, any of which could be selected by the user. Wurlitzer's sales of only 5,000 machines in 1934 jumped to nearly 30,000 in 1939. While not the first firm to manufacture such an automatic player, Wurlitzer was one of the most successful, along with manufacturers such as AMI, Seeburg, and Rock-Ola.

The jukebox not only consumed records but also helped to bring customers back to record stores. Each disc held in the machine was identified on the front panel by title and artist, so that customers could easily find out who they were listening to. The discs themselves were the focus of attention, with manufacturers adopting designs that allowed customers to watch the fascinating process by which the machine automatically selected the correct disc, moved it to the turntable (or, in some cases, moved the turntable to the disc), played it, and then returned it to its place. Jukebox owners tended to own fleets of machines installed in a geographic territory, and sent technicians round regularly to collect the proceeds, change the records (which wore out rapidly), and make sure the machines still worked properly. The constant turnover in records, multiplied by the hundreds of thousands of jukeboxes in service by about 1940, translated into a significant source of sales for the record companies.

REBIRTH

As the 1930s drew to a close, executives in the record industry were publicly talking about the rebirth of the phonograph as a consumer item. World War II began in Europe in 1939, and by late 1941 the Unites States had entered the war as well. Record production was sharply curtailed, and phonograph manufacturers had to retool their facilities to manufacture weapons, radar systems, or other war-related products. Wurlitzer, for example, received government contracts to manufacture gun sights, missiles, and other products completely unrelated to music for the duration of the war. Supplies of shellac, the product used to make most consumer records, were cut off by the Japanese blockades in Asia. To supply records for army camps overseas, people were asked to turn in records to be scrapped (in contrast, in World War I Americans had been asked to donate their used records to be sent overseas and played in the camps). The shellac was extracted from these records and used to make a variety of items. Compounding the record industry's problems, the American Federation of Musicians announced a ban on all recording activities by union members starting in 1942, leading to a severe shortage of content for records. Until these conditions passed, the consumer record and record player industries would remain dormant. The war, however, proved to be an enormous stimulus to the development of new recording technologies.

10

Recording in World War II

THE RECORD OF WAR

Sound recording played a role in World War II in two distinct ways. The first was related to its traditional role in providing entertainment. In this respect, the governments of the world systematically incorporated sound recording into their military support efforts and propaganda campaigns. The second role for recording was related to journalism. The sights and sounds of today's armed conflicts are routinely captured and displayed to the public. Such scrutiny was not always possible. While the war in Vietnam in the 1960s was notable for having its images televised, World War II was the first to have its sounds recorded. New technologies emerged that made portable recording truly practical and started a revolution in "location" journalism. But the stimulus the war provided to the field of sound recording went well beyond that.

THE U.S. ARMED FORCES RADIO NETWORK

Sound recording, which had remained marginal in broadcasting in the United States, became an important part of war-related broadcasts nearly overnight. The Office of War Information (OWI), a new federal agency created in 1942, took responsibility for news and propaganda on behalf of the government, both within the United States and overseas. The OWI is most famous for its various propaganda campaigns, such as the famous *Why We Fight* films. In the history of technology, the OWI is interesting because of the way it used existing technologies in new ways or helped push forward innovation. In the field of recording, it often did so by breaking with established traditions or rules. As a government agency, it was concerned primarily with fulfilling its stated mission in the most effective and least expensive way possible. It did not have to worry about profitability or competition with corporations. So, for example, the reluctance of broadcast networks to use recorded programming, which stemmed partly from their strategy of using live network broadcasts as the competitive basis of their business, was irrelevant to the government's interests. The agency began investigating new ways to use sound recording technology in radio, and it discovered several.

One of the first acts of the OWI was to establish a new network of shortwave broadcast stations. Built and maintained by people drawn from NBC, CBS, the World Broadcasting Network, Crosley Radio, General Electric, Westinghouse, and others, the OWI network operated on the shortwave frequency bands that were widely used in Europe but underutilized in the United States. Like the existing national networks NBC and CBS, the OWI network operated studios in New York City and distributed programming to individual stations located around the country. These included several new stations as well as the existing fourteen-station Voice of America propaganda network. The Voice of America service, which could be heard nearly anywhere in the world, used primarily live broadcasting, but in 1943, the New York City studios began to record archival copies of all broadcasts on 16-inch transcription recorders. Because the VOA broadcasts amounted to international diplomacy, the government prudently decided to keep records of the transmissions. Other government agencies also began using transcription recorders extensively for purposes such as monitoring foreign shortwave broadcasts and telephone conversations. The federal government's attitude toward using sound recording technology for record keeping contrasted with the policies of the networks and local stations, which only irregularly kept such archives (Morton, "History," 1995, 254).

Government broadcasting expanded overseas during the course of the war. Beginning with a handful of stations cobbled together by military technicians, the military began offering entertainment and news from stations at overseas bases and sometimes even near the front beginning in 1942. The new Armed Forces Radio Service (AFRS), with its headquarters in California, began commissioning entertainment recordings to be distributed

Even in World War I, records had been supplied for the entertainment of the troops. By the time of World War II, the U.S. government had formed its own bureau to supply recorded entertainment. Library of Congress, Prints and Photographs Division, Washington, D.C.

to local military stations at bases in the United States, Europe, Africa, and Asia. While some of these programs were original creations, others were simply off-the-air network broadcasts with the commercial announcements deleted. The networks, advertisers, and performers' labor unions agreed to donate this material for the war effort. The recording activities of the AFRS gave a big boost to southern California's transcription record-pressing businesses, which had previously been engaged mainly in recording air-checks and small batches of syndicated radio programs. Between January 1943, when these activities began, and May 1945, well over a million discs were recorded, mastered, and pressed in the Los Angeles area. AFRS activities benefited makers of ordinary phonograph records as well. United States record companies were shipping over 330,000 78 rpm "Victory Discs" to American bases every month by 1945. While these purchases from U.S. companies helped them economically, they also showed how records and transcriptions could provide entertaining radio programming, despite the insistence of the networks that it was inferior. The recording-based, military "networks" would provide a model for postwar radio, which depended much more heavily on recorded content (Morton, "History," 1995, 255; Sanjek 1988, 219).

MILITARY SPONSORSHIP OF MAGNETIC RECORDING

American Telegraphone Company had tried and failed to sell magnetic wire recorders in the early 1900s. The technology then sat, unexploited, in the United States until the late 1930s, when several inventors and corporate laboratories tentatively experimented with new designs. But it was military and federal sponsorship for research, along with purchases of equipment by the government, that had the effect of bringing magnetic recording technology from the laboratory to the market. The Bell Telephone Laboratories, which had earlier successes with "electrical" phonograph recording, motion picture sound, and transcription recorders, first investigated magnetic recording around 1900 and returned to it sporadically over the years. Late in the 1920s, AT&T took note of magnetic wire recorders then being sold in Europe, and decided to develop a similar product in 1930. Undertaken at the newly formed Bell Telephone Laboratories, this development work focused on a steel-tape recorder for possible use within the Bell System. By the mid-1930s, Bell Labs engineers had a machine ready for production, and transferred the project to Western Electric to be manufactured and sold. The recorder used a loop of steel tape to record a few seconds of sound and then play it back immediately. AT&T distributed prototypes of the machines to local telephone operating companies around the United States for their evaluation. These local companies usually ran voice-training classes for newly hired operators, and AT&T believed that the loop recorders would be useful in teaching new employees their jobs. Unfortunately, the marketing effort failed, and by 1940 Bell Labs was looking for new ways to commercialize a technology they had by now spent years developing.

That new market emerged as a series of innovative if obscure devices. The most successful was a gun locator. Using microphones, an amplifier, and a steel-tape magnetic recording device, the gun locator would use a sort of aural triangulation to determine the general location of a distant artillery piece in the battlefield, based on the sound the gun made when it was fired. By adjusting knobs and observing a readout, the operator of the gun locator could get a rough reading on the distance and direction to the gun. The device supported air or ground assaults against enemy gun emplacements. Other, even more remarkable proposals would follow. Western Electric was already selling a powerful public address system, known as "the heater," to the military for use in battlefield deception. The heater blasted battlefield sounds (such as recordings of tanks in motion) toward the enemy in an attempt to create a diversion. A later Bell Laboratories project proposed to fit the heater and a magnetic tape recorder inside a 21-foot-long torpedo. The torpedo would then be fired from a ship or submarine toward the shoreline near territory occupied by the enemy. When it neared the shore, the torpedo would stop automatically, drop anchor, and then float vertically with its nose above the water. The top would pop off, and a loudspeaker would emerge and aim itself at the shore. Then a magnetic wire recorder would play back the sounds of a fake invasion, hopefully deceiving the enemy. This strange device was probably not manufactured in large quantities, and while it was clearly a novel use for magnetic recording, it also reflected the new Market forces propelling the development of magnetic recording.

Military sponsorship also shifted the focus of the Brush Development Company of Cleveland, Ohio. Brush was the corporate descendant of the manufacturing company founded by Charles F. Brush, a pioneer in the field of electric lighting. The Brush Development Company had specialized in the commercial application of a class of mineral crystals called piezoelectrics, which produce voltage when subjected to pressure, or which respond to applied voltages by physically vibrating. Brush had an important series of patents for "crystal" microphones and phonograph pickup cartridges, which were bringing in royalty fees by the late 1930s, and this encouraged them to expand into new areas. One of the fields they chose was magnetic recording.

Brush hired Semi J. Begun, a German-born engineer who had left Nazi Germany in the late 1930s. His experience in Germany included the development of the C. Lorenz steel-tape recorder (discussed later in this chapter), and he now proposed a number of commercial applications for a tape recorder of his own design. After being hired, Begun commenced with a series of experiments in magnetic recording that would result in advanced machines capable of recording sounds on steel wire. Some of these recorders were employed by the Navy and other branches of the military for training and record-keeping purposes to record voice conversations. The wire recorder could also be easily adapted to record radar blips or other electromagnetic pulses other than voice signals. Such recorders were used experimentally during the war to record and reproduce oscilloscope images, apparently with the ultimate aim of recording what operators saw on their radar screens. Radar was one of the most important new technologies of the war, but there was no way to store the images that appeared briefly on the radar screen for later study, except for the awkward and slow technique of simply photographing the screen. In a sense, then, this was the first video recorder, although it was radar blips and not television that was recorded.

By the end of the war, Begun had completed a number of prototype magnetic recorders for this and various other "data"-recording applications. These could record on wires, steel tapes, or metal cylinders. The metal cylinders or drums were important for short-pulse recordings because the constant rotation of the drum allowed a recording to be played over and over again. In this way, the drum recorder could capture a radar blip or other short piece of information and "remember" it, keeping it on the screen so a radar operator could have more time to study it. The drums could record other types of data as well. Although it is beyond the scope of this book, it is worth mentioning that immediately after the end of World War II, Brush drum recorders became the memory devices for several of the earliest computers, and eventually developed into today's computer disc drives. While Brush is rarely mentioned in the history of technology, during World War II its research projects and outright sales to the military brought the company over $18 million in income, a large figure for a such a small company. How much of that was related to magnetic recording is not clear, but at least $1.7 million worth of wire recording equipment was sold to the U.S. Navy.

ARMOUR

OWI sponsorship also brought a little-known university research organization into the international spotlight. Located within what is today the Illinois Institute of Technology in Chicago, the Armour Research Foundation was then the type of small research-for-hire laboratory that nearly every engineering college maintained. It provided local industry with the laboratory facilities and trained personnel that they lacked to conduct product- and process-oriented research. This research translated directly into improved products or more efficient production for manufacturers, and it brought profits to the university. The foundation hired a young electrical engineer named Marvin Camras in 1939, partly because of an undergraduate project in which he designed a magnetic recording device. Convinced that key aspects of the design were patentable, Armour leaders reckoned that with little investment they might license the right to make the recorder to local radio and phonograph manufacturers. At that time, Chicago was second only to the New York–New Jersey area as the center of the electronics industry in the United States.

Between 1939 and 1941, Camras worked on a portable version of his wire recorder suitable for mass production. He found it difficult to improve the sound of the original design without considerable research into the recording/reproducing electromagnets and the recording wire. The wire that he ultimately settled on was a single strand of stainless steel requiring custom manufacturing techniques. Looking beyond the consumer market, Armour gave demonstrations of the recorder to officials from the Army and Navy, and subsequently redesigned the machine to create a rugged military version. The Army Signal Corps was interested in using sound recorders for on-the-spot news reporting, and asked for a battery-operated portable model. American radio journalists who covered the war were usually required to work within the Signal Corps. They were then given access to locations overseas, closer to the front. But the networks' reliance on live news made reporting nearly impossible overseas, not only due to time zone differences but also because of the challenges of establishing the necessary telephone "feeds" to connect the reporters to the studio. In the case of reporter Edgar R. Murrow's famous rooftop broadcasts during the battle of London, reporting that was in effect "behind the lines" could be quite riveting, but it was rare for the network journalists to find such opportunities.

The Signal Corps recognized the potential value of the portable Armour wire recorder for war reporting. The battery-operated version of the machine could be carried by a single person and could be operated while in motion. Unlike disk or optical recorders, magnetic recorders are nearly impervious to sudden changes in temperature, humidity, or the kind of physical shock a recorder is subjected to in the field. After the Armour

NBC "Blue" network reporter George Hicks interviewing a sailor, shortly before the D-Day invasion of Normandy. In the foreground, a technician watches a "Recordograph" sound recorder. The short-lived Recordograph captured sound in a groove like a phonograph, but did so on a strip of plastic rather than a disc. Still Picture Branch, National Archives, Washington, D.C.

recorder proved itself in battle, the Signal Corps began to play up the innovative, mobile use of sound recording in journalism, issuing press releases that were published later in newspapers and magazines. Following the battles of Bougainville and Saipan in the Pacific, for example, several publications ran stories not about the battles themselves, but about the journalistic use of wire recorders to interview soldiers after the fighting or even to capture the sounds of the battle in progress. One reporter strapped the recorder to his body and jumped out of an airplane with a group of paratroopers, recording his impressions of the descent. Such things were simply impractical with any other type of recorder.

Armour sold at least $500,000-worth of these small recorders to the military, producing upwards of several thousand machines of several different designs. At the end of the war, the foundation moved quickly to reintroduce the consumer version of the wire recorder, hoping to capitalize on the publicity the military recorders had received during the war. They found numerous radio and phonograph manufacturers eager to buy licenses that would allow them to manufacture the consumer recorder and receive technical assistance directly from Armour. However, Armour's plans came unraveled when German tape-recording technology was transferred to the United States and other Allied nations. Seizing the Germans' patented tape technology as the spoils of war, manufacturers introduced the modern form of magnetic tape recording by about 1950 (Morton, "Armour Research Foundation," 1998, 213–244).

GERMANY AND THE TAPE RECORDER

Developers of magnetic recording devices in Europe after 1900 had taken a different path than their American counterparts. Following the expiration of Poulsen's European and American patents between 1915 and the early 1920s, a number of new firms entered the business. Partly because of the prominence of German manufacturers in telephone, radio, and phonograph engineering, it was the Germans who led this development. The inventor Curt Stille, an otherwise obscure figure in the history of technology, was one of the most important individuals in this story. Stille made a number of improvements to the original telegraphone and set about licensing the patents to manufacturers. By about 1925, a licensee named Karl Bauer had developed a new machine, perhaps with the help of Stille himself. Bauer's company, Echophon, manufactured and marketed the machine, which was called the Dailygraph, for office dictation purposes. The Dailygraph was a compact, well-designed wire recorder with electronic amplification, suitable for recording directly from the telephone lines, and had a starting price of approximately $800. It also has the technical distinction of being the first commercially available magnetic recorder to house its recording medium in

Women workers winding and packing reels of magnetophone tape at a plant in Germany, circa 1945. U.S. Department of Commerce, Washington, D.C.

a removable cartridge—a distant relative of the tape cassettes of the later twentieth century.

Stille's recorder must have been impressive at the time, because it attracted considerable attention and inspired others to use it as the basis of their own modifications. In the later 1920s, for example, the English movie producer Louis Blattner purchased a license to make the machines, and after substituting flat steel tape for the wire in the original design, he introduced his Blattnerphone for providing motion picture soundtracks. The Blattnerphone was a commercial failure, but it was this machine, modified some more, that was soon reintroduced in the studios of the BBC as the "Marconi-Stille" recorder and was used for broadcasting shortwave programs for many years (Lafferty, "The Blattnerphone," 1983, 18–37).

Meanwhile, Bauer sold the Echophon company to the telephone equipment manufacturer ITT, and the machine's production fell to ITT subsidiary C. Lorenz, which later produced a redesigned Dailygraph that was called the Textophon (usually written Textophone in English). This telephone recorder/dictation machine also used wire as its recording medium.

The Textophone appeared about 1935, and by this time the Nazi party had seized governmental control in Germany. Because of the oppressive nature of the regime, the government stepped up the surveillance of its own population as well as its foreign enemies. In the late twentieth century, stories surfaced about the involvement of the IBM Corporation in supplying automatic record-keeping machines that were used by the German government to keep detailed records of the Jewish segment of the population (Begun 1949, 1–12).[1] Less well known is the use of recording equipment including the Textophone to record telephone conversations of persons under surveillance. According to Semi J. Begun, the engineer who worked for C. Lorenz in the 1930s before coming to America, thousands of Textophones were sold to the government for this purpose, as were the later C. Lorenz tape recorders known as the Stahltonbandmaschine.

The Stahltonbandmaschine (which, roughly translated, means "steel audiotape machine") came into use after 1935, when it was proven especially suited for portable recording. The German broadcasting authority, the RRG, installed the recorders in rolling studios called sound trucks, using them to gather news and entertainment from a variety of locations. Germany, more so than many other nations, also made a point of using public address technology to broadcast entertainment, news, or propaganda to its populace, and the Stahltonbandmaschine excelled in providing a mobile source of programs to be "broadcast" on public address equipment.

The drawbacks of steel-tape recording were clear even at that time. A high tape speed of 1.5 meters per second was necessary to record and reproduce a range of frequencies comparable to the disc-recording equipment of the day, which had an upper limit of 5,000–6000 Hz. At that rate, a reel of tape sufficient to carry a half-hour program was large and heavy, even though the steel tape was only 3 millimeters wide. This meant that a massive and sturdy mechanism was required to start and stop the reels without causing damage to the tape or recorder. The tape itself, like the tape used on the Marconi-Stille recorder in Great Britain, was made of a special type of steel supplied by a Swedish firm. For a country at war, as Germany was after 1939, it was not strategically advisable to be dependent on a foreign supplier.

The C. Lorenz steel-tape recorder was nonetheless widely in use in Germany at the outset of World War II. Like the Marconi-Stille recorder, it was valued for its ability to provide on-location recordings as well as recordings that could be used briefly and then erased easily. It was not long

1. See, for example, Edwin Black's controversial *IBM and the Holocaust* (New York: Crown Publications, 2001).

before a lighter, more compact tape recorder made of all-German components came along. This new machine was the magnetophon (magnetophone in English), designed in the early 1930s and demonstrated in 1935. The magnetophone had advantages due to the fact that it used a tape made of lightweight paper or plastic, coated (or in some cases embedded) with microscopically fine iron powder.

The manufacture of these powders was critical to the system, so the AEG company, the recorder manufacturer, formed a partnership with the chemical firm I. G. Farben. This company became infamous in later years because it also manufactured explosives, synthetic fuel, and the poison gas used to kill prisoners in concentration camps. It was actually a huge conglomerate of firms making a wide range of chemicals including fabric dyes and other industrial chemicals. One of its divisions, BASF, located in Ludwigshafen, Germany, was a leader in the manufacture of fine iron powders, which were then being used by the radio industry and as paint pigments. The AGFA division, a maker of photographic film, also had experience in coating chemicals onto plastic films and contributed to the development process. While a number of different tape designs appeared over the next few years, the most common magnetophone tapes in use were based on a cellulose acetate plastic film, coated with a mixture of a lacquer binder and a carbonyl iron powder. The iron particles were formed through a chemical process rather than being pulverized from larger pieces. In a vat, iron dissolved in acid was chemically manipulated so that the particles "precipitated" or collected at the bottom of the vat. The powder was then removed, dried, and lightly ground. The resulting particles were 10–15 microns in length and had a generally cubic shape. The shape and size of the particles were critical, because as engineers later learned, high-fidelity audio recording on tape depended heavily on the physical characteristics of the iron oxide particles.

MAGNETOPHONES FOR MUSIC

While the majority of magnetophones were used for telephone surveillance, broadcasting, dictation, interrogation, or monitoring enemy broadcasts, there were ongoing experiments with high-fidelity designs, and these would have far-reaching implications. Two improved versions were available by 1945, models HTS and K7, which were competitive with the best transcription disc recorders of the day. The K7 had a frequency range that extended up to 15,000 Hz, higher than most phonographs or radios could reproduce and near the limit of human hearing (which is about 20,000 for people with excellent hearing). These high-quality machines consumed tape at the rate of about 1 meter per second, so that a 1,000-meter reel of tape could hold about 20 minutes of sound. Of approximately 3,300 magnetophones sold between 1936 and 1945, only a handful of these were the high-fidelity HTS models. The components for the advanced K7 model were constructed, but no complete machines were assembled until after the war had ended in 1945. There was even a demonstration of a stereophonic tape recorder, with two half-size heads and a second amplifier channel connected to a standard machine. However, this was apparently used only experimentally.

Even before the HTS machines became available, the magnetophone was praised in Germany. One engineer who witnessed the first demonstrations in 1935 wrote, "Dealers and other people interested literally stormed the demonstration room, so that we were forced to close the door temporarily." The following year, AEG invited Sir Thomas Beecham and the London Philharmonic Orchestra to make a series of recordings during their European tour. Bringing the entire group to Ludwigshafen, AEG captured some of the earliest symphonic recordings ever made on tape. By most accounts, the recordings were not very good, but AEG continually improved the performance of the magnetophone in later years. The RRG placed its first orders for an improved magnetophone, the K4, in 1938, and was apparently using tape recording exclusively by 1942. As the Germans captured much of Western Europe, they brought the magnetophone with them, using it to reestablish the broadcasting services of captured nations.

RECORDING AT THE CLOSE OF THE WAR

Sound recorders and the recording industry had seen great changes by 1945. It took a world war to bring the world economy and the recording industry out of the slump it had been in for nearly a decade. In the United States, the war's stimulus to the recording industry was considerable, as the government placed orders for transcription and phonograph records and recorders, boosting the market for recordings. Wartime concerns shook up the way broadcasting was being done, making it suddenly acceptable to air records and recorded programs. Further, the unique demands of wartime journalism, and the willingness of the military to experiment with new technologies, led to the introduction of portable wire recorders and other new sound-recording devices. At the same time, the Germans (and to a lesser extent the British) had made enormous strides in developing improved tape recorders. Used for a variety of purposes including broadcasting, note taking, and surveillance, this line of development culminated in the magnetophone line of recorders. Even before the war began to turn against Germany, the Allies were curious about these new machines. It remained for them to capture, study, copy, and begin to use this new type of tape recorder.

11

The Postwar Scene

THE FIAT PROGRAM IN GERMANY

Even before the war in Europe was over, British and American intelligence agencies were shifting some of their attention from the war at hand to postwar strategies. Intelligence agents from the beginning of the war had to put much of their effort into identifying German military and industrial sites. The latter included the huge chemical complex at Ludwigshafen, where I. G. Farben had made the synthetic fuel on which the German war effort depended heavily. Yet nestled within the enormous complex was also the small magnetic tape-making facility of BASF. While in earlier years, agents had tried to determine what parts of the I. G. Farben complex at Ludwigshafen should be targeted for strategic bombing, now they were more interested in identifying technical and scientific "targets" to visit in order to gather information. In August 1944, the Allied Joint Chiefs of Staff established a Combined Intelligence Objective Sub-Committee to manage the process of technical and scientific information gathering. New technical intelligence groups called T-Forces would be attached to infantry groups and would be responsible for identifying, cataloging, and if necessary transporting any strategically valuable technologies they found. British and American T-Force members were drawn from both military and civilian personnel; the civilians were volunteers recruited from the chemical, textile, metalworking, ordnance, and electronics industries.

PROJECT PAPERCLIP

After the end of the war in 1945, intelligence-gathering individuals and groups were spread out over Germany. Teams of investigators visited a wide range of German military installations and manufacturing facilities, looking for anything of interest. The groups had the authority to seize corporate documents and property, and to interrogate virtually anyone deemed knowledgeable. British and American teams were often investigating the same sites in competition, resulting in confusion. Because the Russians were also conducting the same kinds of investigations, the British and Americans rushed to get the best information first. There were two major areas of technical investigation, the first focused on German technologies of war, particularly in the areas of aerospace and electronics, and the second aimed at plundering German technological advances in nonmilitary fields. The military investigations had priority, since in theory the resulting intelligence could contribute to the ongoing war against Japan. Always in the background, however, was a growing rivalry with the Soviet Union over this technology. Soviet and American investigators were actively seeking out German engineers and scientists in key fields, attempting to lure them over to their side with attractive offers. The U.S. government authorized the military to offer employment in the United States to scientists identified as having key knowledge of certain military technologies. The most famous example of this activity was the case of the German "rocket scientists," especially Werner Von Braun. Von Braun and others had designed the famous V2 rockets used mainly against England and Belgium during the war. A group of these rocket scientists was relocated to a military base in Alabama, where they contributed to both the missile buildup and the space race in the postwar period. The Soviets did much the same thing in their country.

The collection and distribution of technical intelligence were later reorganized under a new administration, the Field Intelligence Agency Technical (FIAT), in 1945. The FIAT program was clearly less successful at transferring German industrial knowledge than had been Project Paperclip. It appears investigators found few examples of German industrial technologies that were significantly different than what the Americans or British already had. An important exception was the German tape recorder.

The lead FIAT investigator in tape-recording technology was Signal Corps Colonel Richard H. Ranger, whose previous career included pioneering work in the field of wireless facsimile transmission for RCA. After being stationed in the United States during the early part of the war, in mid-1944 Ranger was reassigned to Europe and led technical intelligence missions in England, France, and Germany. In late 1944 or early 1945, he began his investigation of the magnetophone, which involved reviewing some of the related AEG corporate documents and interviewing key personnel.

By early 1945, Radio Luxembourg and other stations in the area were being put back on the air. While there were many serviceable magnetophones available, tape was in short supply. Investigators learned that the Ludwigshafen tape-manufacturing facility had been damaged by bombing in 1943. After that, only oxides and tape base were made at Ludwigshafen, while coating was done at an AGFA plant in Wolfen, Germany. John H. Orr, a civilian working with the OWI, was in charge of starting tape production. Orr was a self-taught radio serviceman from Alabama who had been handed the tape project late in 1945 and by December had a tiny production line set up in Wald Michelbach, a village in the American zone of Germany.

Ranger and Orr discussed the possibility of going into business after the war, and both made arrangements to send magnetophones and samples of tape back to the United States. Before they had made much progress, Orr was sent back to the States, apparently because of an injury. Ranger in late 1945 also returned home and set to work designing a recorder- and a tape-production line. His Rangertone recorder, publicly demonstrated in 1947 but probably not in regular production until the next year, was somewhat different in design than the magnetophone. Yet Ranger had adopted all the magnetophone's major operating features, optimizing the Rangertone for use with magnetophone tape, running the tape at virtually the same speed of 30 inches per second (rather than 1 meter per second), establishing the supply reel as the one on the left-hand side, winding the tape with the oxide facing inward, and so on. The most critical parts of the recorder, the electromagnetic "heads," were virtual copies of magnetophone heads. Yet Ranger made no attempt to hide the similarities between the Rangertone and the magnetophone, and in fact he frequently acknowledged his debt to the German engineers. Similarly, John H. Orr never concealed the origins of his technology. After all, the magnetophone was officially no longer German property. After returning to Alabama, Orr obtained the tape-coating machinery partially completed by Richard Ranger. Establishing his factory in the small town of Opelika, Alabama,

Orr proudly announced the manufacture of his first reels of Orradio tape in early 1949. This tape, coated on "kraft" paper (the heavy, brown paper used to make paper bags), was as close a copy of the German tape as Orr could muster.

Yet it was clear that hands-on experience in Germany was not a prerequisite to making tape recorders or tape, as some relied largely on FIAT for their information. Late in 1945, the U.S. Department of Commerce took over the republication of the declassified FIAT technical reports, making them available to the public and advertising their availability. Anticipating a market for recording tape, the Audio Devices Corporation of New York (previously a maker of recording discs) started working toward magnetic tape production based on Department of Commerce publications. The company introduced its Audiotape brand in 1949.

While Rangertone, Audio Devices, Orradio Industries, and others would attain success in the recorder or tape fields in the 1950s, two other firms stand out not only for their products but also for the ways they entered the business. One, Ampex Corporation, became the premier maker of professional tape recorders. The other, the Minnesota Mining and Manufacturing (3M) Company, was for many years the world's largest manufacturer of recording tapes.

The Ampex Electric Corporation was an electric motor manufacturer looking for a new product in 1945. The company hired engineer Harold Lindsay, who attended a demonstration of the magnetophone given by former FIAT investigator Jack Mullin. Like Orr and Ranger, Mullin had gained firsthand experience with magnetic recording technology in Germany, and had managed to ship two recorders and samples of tape back to the United States. With a partner, Mullin demonstrated the machine to the San Francisco section of the Institute of Radio Engineers in 1946. Lindsay, who attended the demonstration, later convinced Ampex founder Alex M. Poniatoff to build a similar machine with Mullin's assistance.

Yet the subsequent appearance of the first Ampex Model 200 in 1948 was not all attributable to German technology. Ampex took a license from the Armour Research Foundation in Chicago, which gave the company access to Armour's wide-ranging experience in making magnetic recording devices. Armour by this time had also learned about the magnetophone, and the foundation's lead researcher, Marvin Camras, had hastily designed a tape recorder in late 1945. While the Ampex had much less in common with the Armour machine than the magnetophone, Armour's base of knowledge must have been helpful to Ampex as company engineers learned about the unfamiliar technology of magnetic recording.

The mixing of American and German ideas is also evident in the way Minnesota Mining and Manufacturing entered the tape business. Back before the war, in 1939, Brush Development Company had unsuccessfully introduced an endless-loop, steel-tape recorder called the Soundmirror. It was designed by Semi J. Begun and based on his knowledge of steel-tape technology in Germany. Begun later became interested in German-style powder-coated tapes, and Brush introduced a home tape recorder also called the Soundmirror in 1945, before Ampex or Rangertone had gotten their machines into production. Brush also licensed the Amplifier Corporation of America, a long-forgotten firm in the radio field, to make a version of the recorder, which it did under the trade name Magnephone beginning in 1946. Brush turned to others to produce the recording medium, which consisted of a kraft-paper tape coated with the same powdered iron oxide used by the Germans. Of several candidates, 3M was the leading manufacturer of a similar product, an adhesive tape marketed as Scotch Tape. While the oxide used for Brush tape had slightly different magnetic properties than the latest magnetophone tapes, it would work well at the Soundmirror's lower tape speed, and in 1946 and 1947 the company began supplying small quantities of its "Type 100" paper-based tape to users of the Soundmirror, introduced that year, and the captured German magnetophones now in the United States.

Later, Armour and Brush independently experimented with a new type of oxide for making tape, and after Armour specified this oxide in its purchase orders for tape (apparently to the DuPont Corporation rather than 3M), it became the new standard formulation. This new oxide had an acicular, or stick-like, shape rather than being cubical, and it had slightly different magnetic properties. Its chief advantage was that it worked better than the cubic oxide at the low tape speeds that both Brush and Armour were working toward. Prototype Armour machines operated at a tape speed of 8 inches per second, and Armour claimed that it had discovered the new oxide on its own (leading to a bitter fight in court in the late 1950s). Once the new oxide was available, Ampex and other makers of studio recorders offered machines that operated at multiple speeds: 30 inches per second for maximum quality, 15 inches for voice recordings, or even 7.5 inches for rough work. When consumer recorders began appearing in greater numbers around 1950, most of them operated at 7.5 inches per second, or even 3.75 inches per second. Paper-backed tapes were soon joined by Scotch type 111 tape in 1948, which had a stronger, smoother, acetate plastic-based tape and used the new oxide. Scotch 111 quickly became the top seller in the United States, and remained in use for many years—into the 1970s in the case of consumer tape recording.

TAPE IN BROADCASTING

The radio business was already primed for the transition to the use of tape. Following the famous *Hindenburg* broadcast of 1937, networks and local stations began using recorders more frequently for capturing "spot" news at remote locations. In Europe, where there was no tradition against using recorded programs, radio stations adopted tape without controversy. The heavily publicized use of portable Armour wire recorders by the U.S. Army Signal Corps probably encouraged their use among American radio journalists. As early as 1946, stories began to appear of local stations broadcasting news material from wire. WOR in Washington, D.C., for example, broadcast a speech by U.S. Interior Secretary Harold Ickes in February 1946, and Chicago's station WMAQ began a series of wire-recorded news programs later that year.

A thirteen-part series that premiered in 1946, featuring journalist Norman Corwin interviewing people around the world, was something of a breakthrough because it was probably the first ongoing series of recorded news features aired on a network, the American Broadcasting Corporation. ABC had emerged earlier as the third U.S. network as a result of the court-ordered breakup of NBC in 1943, and was apparently inclined to break old network traditions if necessary to gain ratings. Corwin's series was followed in 1946 by CBS's pathbreaking use of a Brush model BK-401 tape recorder for news broadcasts. These recordings, consisting primarily of excerpts from the 1946 Republican and Democratic conventions, were novel in that the tapes were edited to condense their content (editing was so time-consuming with discs that it was rarely undertaken, although it was a standard part of motion picture production).

Historian William Lafferty has also suggested that ABC was pushed toward the greater use of sound recordings by the daylight saving time system. Daylight saving time was mandated around the country from 1942 to 1945, but after that local governments were free to stay on it or abandon it. The irregularities in timekeeping around the country and the propensity of municipalities to jump into our out of the system at will created problems for network broadcasters when they tried to publish program schedules.

The major U.S. networks, centered in the East and Midwest, had always faced a problem serving California and the West Coast, where the differences in time put early morning and evening broadcasts at inconvenient hours. For a while, the networks had broadcast their programming twice—once for the East and Midwest, and then later for the West Coast. The enormous expense of assembling the casts and crews twice for every program must have been a strong incentive to use recordings for the second broadcast, but until the advent of instantaneous recording discs, this was probably impractical. It simply took too long to master and duplicate ordinary transcription recordings. Yet this time delay problem, coupled with a journalistic enthusiasm for the wire recorder and the appearance of the magnetophone, was soon to push all the networks toward the regular use of recording technology and in particular tape.

In addition to using wire recorders for news, ABC began to use transcriptions beginning in 1946 to time-delay broadcasts to the West Coast. The recordings were made off the incoming land line at WENR in Chicago, and then fed by another land line to Western affiliates at the appropriate hours. The fourth largest network, Mutual Broadcasting System, followed suit the same year, and CBS did the same in 1947.

A second set of circumstances led to ABC's pioneering use of the magnetophone to create a program that was never broadcast live at all. In 1944, ABC had lured away radio superstar Bing Crosby from NBC. Before agreeing to the move, Crosby insisted that he be allowed to record his performances on disc. Crosby's first recorded shows on disc aired in late 1946. However, ABC officials seemed unsatisfied with the sound quality of the discs. The problem seems to have been due to the fact that Crosby rarely gave a perfect performance. He would record several takes, parts of which would then have to be combined on a new recording. Editing discs at that time involved capturing the needed bits from a disc on the fly while cutting a second disc. In addition to requiring split-second timing, disc-to-disc editing resulted in noticeable loss of quality, because the background noise of each generation of recording was added to the next generation.

Jack Mullin was invited to the ABC studio in Hollywood to make an experimental taping of the first Crosby show for the 1947–1948 season. After the show, he stayed in the studio to edit the tape down to its final form. ABC engineers were so impressed that they hired him to tape the rest of the shows. By the middle of the season, Mullin was short on tape, and his recorders were getting worn out. Fortunately, the Ampex Corporation came to the rescue, delivering the company's first two production machines to ABC studios and later supplying a dozen more. What little remained of network resistance to recording technology crumbled almost instantly with the appearance of the Ampex machines. The company's next two models, the 300 and 350/351 series, became standard equipment both in the network studios and at local stations following the model 300's introduction in July 1949. These machines were produced for years, and many were in regular use into the 1980s.

What ensured the ongoing success of tape in the radio business had much to do with the advent of television. Radio networks rapidly transferred their assets, including their best performers, to television beginning in 1947. The radio networks hung on for a few years, but then tended to dissolve, leaving local stations to fend for themselves. Desperate for low-cost content, local stations began using recordings heavily, including not only phonograph records but also news, advertisements, and other programs on tape.

MAGNETIC RECORDING IN MOTION PICTURES

The effort to use magnetic recording devices to provide sound for motion pictures is almost as old as the motion picture itself. However, the technical deficiencies of early magnetic recording technology gave ample room for optical sound recording to become the established method. Even Valdemar Poulsen turned to optical sound recording by the 1920s. In the mid-1940s, Semi J. Begun of Brush Development returned to the idea of using magnetic recording for motion pictures and began promoting in the technical press. About the same time, Marvin Camras also discussed magnetic motion picture soundtracks with Armour licensees, although he seemed most interested in establishing this as a consumer "home movie" technology rather than selling it to Hollywood. Semi Begun in 1946 submitted a formal query to Loren Ryder of the Motion Picture Academy's Basic Sound Committee, asking for the committee's opinion on whether magnetic recording could become the basis of a new studio technology. The query resulted in a list of twenty-one recommendations. The committee clearly felt that for magnetic recording to succeed in Hollywood, it would have to meet or exceed the technical performance of existing optical sound systems, not demand radically different production or editing techniques, and be inexpensive to purchase and use. Ryder, who was director of Sound Recording at the Paramount Studios and had witnessed demonstrations of the magnetophone while serving with the Signal Corps, had become an ardent supporter of magnetic recording and promoted it to his colleagues. By late 1946, Marvin Camras had developed an experimental sound recorder for the motion picture industry that used a filmlike, 35mm sprocketed plastic tape, coated with magnetic iron oxide across its full width.

Some studio executives believed that magnetic recording would provide an important savings in production costs. The cost of the tape would clearly be lower than optical sound recording, which required the film stock to make the negative, processing to develop the film, and the making of one or more positive prints. Ryder strengthened the argument for tape by producing a

A recording studio equipped with RCA's first magnetic tape recorder, circa 1949. David Sarnoff Library, Princeton, New Jersey.

more detailed study showing how magnetic tape's cost savings would also be spread throughout the complex recording and editing process. According to Ryder, the savings would amount to about 82 percent.

Movie producers had to wait until about 1948 to get their hands on the first production model recorders. One of the first on the market was a Rangertone recorder, modified with circuits that allowed it to be synchronized accurately to a motion picture projector. The Rangertone's use of standard ¼-inch-wide, magnetophone-style tape did not prove to be as popular as recorders that used sprocketed tape that could be mechanically coupled to a camera. This was the way that optical soundtracks were usually made by the 1950s.

The existing optical recording equipment manufacturers, such as Western Electric and RCA, moved quickly to offer conversion kits that allowed users to begin running 35mm sprocketed tape. RCA had approached the DuPont Corporation with the idea of coating 35mm film stock as early as 1946, and by 1948 several model PR-23 optical recorders had been converted to magnetic recording and were in active use. Just a year later, nearly every major studio had purchased these conversion kits, which by that time had been supplemented by completely new recording systems offered by RCA Photophone, Westrex, and others.

Magnetic recording was not without its critics. Many editors in the early 1950s refused to work with it, complaining that unlike optical recordings, it was impossible to edit the film visually by looking at the undulations on the sound print. These editors insisted on making optical prints of a magnetically recorded original and editing from the prints, although this nearly negated the cost advantage of working directly with magnetic tape. Paramount engineers actually added what they called a "modulation scribe" to the magnetic recorders, which was simply a ball-point pen driven by an electromagnet, which drew the waveform on the surface of the magnetic tape to give editors a way to edit visually. In response to ongoing objections, manufacturers in the early 1950s developed the "electro-printing" method of making low-cost optical recordings, and while more expensive than magnetic tape, they were less expensive than the older methods of transferring a soundtrack negative original to a positive print used for editing.

Gradually, though, the use of magnetic recording crept into the moviemaking process. Libraries of sound effects at Paramount were, for example, dubbed from optical to magnetic film in 1950. The magnetic versions stood up to frequent use better than the optical prints. Making films "on location," which often meant out-of-doors, was also an area where the advantages of rugged, portable magnetic recording equipment were hard to deny. By about 1950, more location film production was being recorded on tape by technicians working for Universal, Twentieth Century Fox, and elsewhere. The new medium of television also adopted magnetic recording, using electrically synchronized, ¼-inch tape recorders to record sound tracks for use with inexpensive 16mm film. Multitrack recording on 35mm sprocketed magnetic tape was also possible using commercially available machines after 1950, when RCA-Photophone introduced such a recorder. Recording multiple tracks on a single tape was an economical alternative to making multiple, separate recordings on photographic film. Later, however, studios realized that by recording the voice dialog on a track separate from music, sound effects, or other audio material, magnetic tape made it easier to produce different editions of the film, such as versions in languages other than English. Because the American studios distributed films worldwide, any technology that reduced the cost of "dubbing" to another language was valuable. By about 1950, then, much of the original, "production" recording was done on magnetic tape, while much of the later editing was done on film, although nearly all prints distributed to theaters still had optical sound.

THE IMPACT OF TAPE

German magnetophone technology, grabbed by British, American, French, and Soviet interests, was in the 1940s seen as a legitimate spoil of war. In retrospect, since Germany was not required by treaty to pay war reparations, the move looks more like a corporate free-for-all. However this transfer of technology is interpreted, it is clear that proprietary German knowledge about tape recording was widely dispersed and later highly influential. The first postwar tape recorders produced in many countries around the world were little more than copies of the magnetophone. In Germany, the magnetophone would survive and be reintroduced as a profitable line of professional and consumer products by the 1950s. It is unlikely that the AEG/I. G. Farben group could have been as successful in convincing so many makers and users of recording technology to adopt tape, given the momentum of existing disc and optical recording. The postwar plundering was if nothing else genuinely effective in diffusing tape recording worldwide.

The introduction of tape recording was loudly trumpeted at the time and was well underway by 1950, but it had yet to make much of an impact. In broadcasting, users of tape had overturned longstanding objections to the use of recordings, but from the listener's perspective, little had changed. Radio programming went on much as before. Similarly, the motion picture industry was gradually adopting the use of tape for making original soundtracks. While tape was saving the studios money, it was not yet making much of a difference in the way sound studio personnel made or edited recordings, nor was tape making much of a difference in the sounds that moviegoers heard in the theaters. Within a decade, however, the use of tape had utterly transformed the making of original recordings in nearly every context, and was well on its way to becoming an important consumer product.

12

Hi-Fi

REARRANGING THE LIVING ROOM

Before World War II, sound recording had become a worldwide success, but in 1945 the world was in a sorry state of disrepair and the sound recording industry was changing. The decade and a half from the mid-1940s through the end of the 1950s was a time of rebuilding and reconstruction after the devastation of World War II. In what was being called the "third world," longstanding colonial relationships between Europe and Africa, Asia, and the Middle East were undergoing fundamental shifts. West Germany, consisting of the areas formerly occupied by the Americans, the British, and the French, was creating an "economic miracle," while its Soviet-controlled neighbor, East Germany, was becoming more isolated from Europe. Britain was entering a long period remembered as "austerity," with cutbacks and even rationing that made life bleak and gray for many. Japan was busy re-creating a devastated industrial base.

The United States was virtually the only country that was stronger at the end of the war than it had been at the beginning, and this certainly helped it consolidate its leadership of the global sound recording market. In the United States, consumer goods in the late 1940s were sometimes in short supply, not usually because of economic weakness, but because consumers were buying so fast that industry could not keep up with demand. Further, the market for many kinds of consumer goods had fundamentally changed. The country had been transformed in important ways during the period since the late 1920s, when that last period of prosperity had ended. Much more than half the population now lived in urban areas and had easy access to a wider variety of goods and services. Automobile ownership had risen, and automakers famously (perhaps notoriously) offered larger, heavier, and more accessory-laden automobiles in a frenzy of excess that peaked about 1960. The focus of consumption was the home, where so much of the consumer dollar was spent, and it had changed as well. Suburban living was becoming the norm rather than the exception. New houses in the suburbs were built according to updated building codes that, among other things, specified a larger number of electrical outlets and demanded electric lighting in virtually every room. In fact, millions did not have household electrical service until the late 1940s, and it was only after World War II that the majority of the population had telephone service, so now huge numbers of Americans were able to participate in the explosion of electrical and electronic consumer goods. It was in this context of global change and local prosperity that new forms of sound recording emerged.

HIGH FIDELITY

By far the most important of those new consumer goods, both in economic terms and in terms of its impact on society, was television. Television had been demonstrated by RCA in the late 1930s, and stations were on the air in a few cities when the war broke out. But it was only after World War II that inexpensive consumer receivers were available, and it was then that nearly every community in the country got its first television stations. Competing with television for the consumer's dollar was a range of other entertainment technologies, including the traditional radios and phonographs. Yet there was a shift in consumer behavior that tended to lead consumers toward more expensive, more elaborate radios and phonographs than had been the norm in the economically leaner years of the 1930s and 1940s. Consisting of equal parts of marketing hype and genuine technological innovation, this trend was difficult to define but easy to name: high fidelity.

The ownership and use of high-fidelity electronic gear began as the hobby of a band of technological enthusiasts and turned into a widespread fad in the 1950s, before becoming a part of worldwide consumer culture from the 1960s on. Its roots stretched back to the 1930s and the introduction of improved radio receivers and electronically amplified phonographs.

Underlying the phenomenon was an enthusiasm for recording technology. In general, what historians refer to as technological enthusiasm is such a strong feature of American culture that it is remarkable today when technology is actually criticized or rejected. The belief that, in some sense, a better life (or better sound) is achievable through the use of the latest technology is the necessary driving force behind the high-fidelity movement.

The emergence of the high-fidelity movement marked the disappearance of an earlier, more negative set of attitudes toward the phonograph. Some early critics labeled the phonograph a second-rate way to listen to music. Even after Edison's Tone Tests flaunted the improving state of recording techniques, it took years for music lovers to accept the record player (or the radio or motion pictures) as a way to have a fully satisfying music-listening experience. But by the 1930s, as the technology of recording and reproduction was changing rapidly, more people began to see the phonograph's potential for delivering such satisfaction. These people invented the high-fidelity movement.

Who were they? During the 1930s, electrical recording and reproduction and electronic amplification techniques were maturing. The wealthy motion picture and radio industries had fostered the development of greatly improved recording and reproduction devices for their own use. The first high-fidelity enthusiasts were often radio or studio engineers with a love of music and access to the latest technologies. These people played a dual role, ensuring that the material broadcast (or recorded) was of the highest quality possible, but also sometimes promoting technical change in the receivers or record players available to consumers, so that ordinary people could appreciate the high-quality sound becoming available. They did so by publishing "how-to" articles in the many technology-oriented magazines of the day, and sometimes by manufacturing and selling their own equipment. A scientific mindset among technicians and engineers contributed to the notion that the music heard in the home on the radio or on disc should ideally sound exactly like the original performance. It should, in other words, exhibit a high degree of "truth," or fidelity, to the original. And, according to science, that fidelity ought to be measurable. Measuring every aspect of sound was impossible, but engineers of the 1930s and 1940s could measure frequency range, total distortion of the signal, the ratio of signal to noise, and certain other key factors. The ideal recording medium could capture the entire range of frequencies audible to humans, with minimal distortion of the signal, and without adding any extraneous hiss, hum, or other noises. Of course, no recording medium could do that perfectly, but widening the frequency range and eliminating distortion and background noise became the main goals of the movement.

The first use of the term "high fidelity" was in the late 1920s or early 1930s, and probably occurred in Great Britain before it was adopted in the United States. Radio stations at that time were improving their facilities to update the first generation of equipment from the 1920s, and advertised the availability of content with a wider range of audible frequencies than was possible before. For example, electrical engineer John Hogan put station W2XR on the air in New York City in 1935. The station was classified as "experimental" because of Hogan's use of equipment that allowed the transmission of a wider band of frequencies. The station was a local success and became a regular commercial enterprise in 1937, changing its call letters to WQXR. Hogan was so concerned with fidelity because he wanted to broadcast what was then called "good music," that is, classical and other "highbrow" forms of music. This important connection between high fidelity and high culture was typical of the early high-fidelity movement. It was shared by inventor Edwin H. Armstrong, who in 1936 established the first commercial FM radio station in New Jersey. Broadcasting nearly static-free audio to listeners across the river in nearby New York City, Armstrong believed that the best way to deliver high-quality musical content was to deliver it in high fidelity. He lobbied the Federal Communications Commission and the radio manufacturers to get FM stations on the air, and also made sure that the bandwidth allocated to each radio channel was wider than what had been established for AM radio. This helped ensure that FM sounded better to listeners.

The small but loyal group of devotees to high-fidelity sound in the late 1930s was distracted for several years by the onset of World War II, but the war would have the effect of expanding the high-fidelity movement. While there were some improvements in civilian radio during the war, notably the introduction of better quality tape recorders in Germany and elsewhere, most countries halted the production of home radio receivers in order to devote resources to military production. The military services in all countries sought out those with expertise in electronics, and trained thousands of men and women the basics of electrical engineering. In the United States, there were also military men, probably numbering in the hundreds, who gained firsthand experience working on a daily basis with audio devices such as radios, amplifiers, and even recorders. After the war, these people, most of them young men, often pursued their interests in electronics by assembling their own televisions, receivers, amplifiers, and loudspeaker enclosures. Their numbers greatly expanded the high-fidelity community of the late 1940s.

The evidence for an explosion of interest in audio electronics and high fidelity is clear in the numerous technical publications of the day.

Even before the war ended, articles in these magazines were turning from military to civilian concerns. Highly technical periodicals such as *Radio-Television News*, *FM*, *Audio Engineering*, and *Electronics* published scores of articles on audio in these years, giving detailed instructions on how advanced high-fidelity equipment worked and how to build it. While before the war, the focus of high fidelity had usually been limited to home phonographs and radios, after the war enthusiasts began thinking in terms of more complex systems consisting of "components," such as loudspeakers, power supplies, preamplifiers, and power amplifiers. They began building "tuners," which were radio receivers that had to be connected to an external amplifier and loudspeaker. They did the same with the phonograph, purchasing "turntables" in the form of bare chassis that had to be installed in a custom cabinet and connected to an external amplifier. While the details varied considerably, in general the necessary equipment for a postwar high-fidelity "system" by 1950 included at minimum a loudspeaker, an amplifier, and two sources of music: an AM tuner and a turntable.

HOME WIRE RECORDERS

There were many other products for audio enthusiasts to consider, such as FM tuners. FM radios had a limited appeal in the 1930s when they were introduced, because so few cities had FM stations. In the postwar period, the number of FM stations on the air increased rapidly. Television also had high-fidelity potential, because the sound portion of television was broadcast in wideband FM. As more electronics firms returned to civilian production in 1946, many of them began to offer a surprising new product: the magnetic wire recorder. The Armour Research Foundation had enticed dozens of firms in the United States and Europe to become licensees for its patented wire-recording technology. In return for a modest initial fee and royalty payments on the sales of recorders, Armour provided detailed plans for manufacturing a simple, inexpensive consumer recorder. A few firms created their own designs, but most simply offered the standard Armour model. From late 1946 through 1948, tens of thousands of wire recorders were sold. However, this technology was doomed.

Brush Development Company reintroduced its Soundmirror line of consumer tape recorders in late 1946, followed by the 1947 introduction of a similar machine by a Brush licensee, the Amplifier Corporation of America. Then a flood of consumer tape recorders began to appear in 1949, with relatively low-cost, open-reel recorders offered by a host of mostly forgotten manufacturers such as Pentron, Webster-Chicago, Revere, Ferrograph, and others. The first consumer tape recorders had few advantages over the wire recorders in terms of measurable audio performance, yet they quickly stole the entire market, with wire recorders disappearing by about 1954.

The reasons for the wire recorder's rapid demise and the tape recorder's equally rapid ascent had much to do with the way these devices were promoted to the public. As the high-fidelity movement was picking up steam after 1949, articles about hi-fi began appearing in newspapers and popular magazines. Reviews of recordings, which had been uncommon in the 1930s, began to reappear in popular literature, such as E. T. Canby's reviews for *The Saturday Review of Literature*, which began in 1945. The hi-fi hobby was big enough in 1950 to support the publication of *High Fidelity Magazine*, a journal for recorded music lovers that was published until 1988. In *High Fidelity* and elsewhere, critics and purists used the power of the printed word to influence the tastes of hobbyists. These critics saw the tape recorder as the more advanced, and hence more desirable, technology, probably because they were familiar with the high-quality studio tape recorders then coming into use. To them, the wire recorder was not worth serious consideration no matter what its capabilities or potential.

Yet even the endorsement of the most prominent high-fidelity boosters was not enough to make the tape recorder an essential part of the home "hi-fi" system. Statistics for the United States, which had by far the largest market for tape recorders, show that hundreds of thousands of consumers bought tape recorders in the 1950s. Yet such impressive sales were dwarfed by the many *millions* of sales of the familiar phonograph.

HIGH FIDELITY AND THE DISC

The first postwar phonographs were little more than updated versions of the record players of the 1930s. The 78 rpm record remained in production, and standards of fidelity for consumer discs were unchanged by the introduction of tape in the studio. Within a few years, however, the publicity generated by the high-fidelity movement was building a new market for more expensive phonographs with the features desired by the hobbyists. High-fidelity enthusiasts could not build their own phonographs as easily as they built their own amplifiers and loudspeaker enclosures, but they could compare products and select the most appealing turntable, tonearm, and pickup combinations. Expert reviewers in magazines directed buyers toward phonographs that operated with only minor variations in speed, and transmitted little "rumble" to the sensitive pickup. The tonearms they preferred were the lightweight models then being offered on some phonographs, because these put less unnecessary weight on the stylus. The low weight was better for the new generation of lightweight pickups, such as the General Electric "variable reluctance" models, or the numerous brands based on the Brush Development Company's patented piezoelectric technology, all of which had better frequency response than their prewar predecessors.

High-fidelity players appeared before any major improvement in the records to be played on them. Building on innovations in cutting-head technology developed in the 1930s but not yet implemented, engineers Arthur Haddy and Kenneth Wilkinson of the Decca Record Company developed an improved, "moving-coil" disc-cutting head in the early 1940s. Its original purpose, related to the war effort, was to make wide-range recordings of the noises generated underwater by submarine motors. In 1944, Decca used this same technology to make its first high-fidelity recordings on 78 rpm discs. Known by the less-than-catchy name of Full Frequency Range Recordings (FFRR), the improved Decca discs were an instant hit not only in Britain but also in the United States.

BACKGROUND OF THE LP AND THE 45 RPM DISC

The FFRR recordings and other minor variations (such as special, 100 percent-vinyl discs) sustained the growing high-fidelity movement for several years, but it was the introduction of the Columbia long-playing disc that propelled hi-fi into the 1950s. The LP, as it was known, incorporated many of the improvements in disc recording that had been available but idle since the 1930s. It used a finer, narrower groove than the old 78 rpm disc, which demanded a more sensitive pickup and a virtually microscopic stylus tip, but which made wide-range recordings more practical. The discs themselves were 12 inches in diameter to hold a longer recording, and were made of vinyl plastic rather than the grit-filled shellac compound used since Berliner's day. The vinyl, combined with cleaner recording rooms and hi-fi playback equipment, produced recordings with much less background noise.

The new Columbia LP reflected not only technical concerns but also the interests of classical music listeners, particularly the project's director, an engineer and manager named Peter Goldmark. The disc's 12-inch diameter, finer groove, and slower 33⅓ speed added up to a record that could easily hold about 20 minutes of music per side. Columbia engineer Edward Wallerstein had earlier calculated that with a 20-minute playing time, the

RCA's 45 rpm disc, launched in the late 1940s, was intended to be a high-fidelity replacement for the old 78 rpm record. Its key feature was a new record changer. Shown to the side are "albums" of discs. David Sarnoff Library, Princeton, New Jersey.

discs could hold nearly any movement of existing Columbia classical recordings on a single side. Many other engineers participated in the LP's development. William Bachman, for example, was brought into the project in 1947, and contributed an improved record cutter that used a heated stylus to slice smoothly into the master disc. Another CBS engineer named Rene Snepvangers contributed to the development of a lightweight pickup, which CBS would later include in the special record players needed to listen to the LP. The development of the disc, which had cost only about $250,000, was seen by Columbia President William S. Paley as a niche-market product that would complement, rather than replace, the ordinary 78 rpm single.

Columbia's hi-fi product of the late 1940s was the long-playing (LP) record. Conceived as a way to present long recordings of classical music, it also stimulated the release of popular albums, and eventually RCA and others began manufacturing LPs as well. David Sarnoff Library, Princeton, New Jersey.

In late 1947, as the project was nearing completion, CBS reached an agreement with the Philco Radio Company (once a major radio manufacturer in the United States) to produce an inexpensive player for the records, since CBS did not have its own player manufacturing facilities. The

CBS record manufacturing plant at Bridgeport, Connecticut, which had been in service for decades, was retrofitted with equipment to handle the LP records. Then, in early 1948, the company held its first public demonstrations. Later that year, in June, CBS organized a press party at the fancy Waldorf Astoria hotel in New York City. Wallerstein and his team made what became a famous demonstration not only of the improved sound of the disc, but also of the fact that an 8-foot-high stack of 78 discs, placed next to a 15-inch stack of LP discs, both contained the same amount of music.

THE 45

While Columbia was developing the LP, RCA-Victor was at work developing the successor to the 78 rpm disc. The project began before World War II, when Victor's share of U.S. record sales was slipping. Engineers within the company proposed a new, smaller record and a compact player called the "X-changer." The new player was a record changer, similar in form to those introduced in the 1930s, which automatically played a series of records stacked up on a spindle mechanism and loaded onto the platter one by one. The records played at a new speed, 45 revolutions per minute, which was chosen partly to gain better sound quality, and partly to allow the changing mechanism to finish its cycle in about 5 seconds. The discs packed about as much music as an ordinary 12-inch 78 rpm disc (about 5 minutes' worth) onto a smaller, 7-inch disc by using a finer groove—about the same 1-mil groove width that CBS had selected for the LP. In fact, with minor modifications, the LP and the 45 could be played by the same machine. The major difference between the two, besides the size and speed, was the hole in the center. For the LP, CBS had used the same small hole as the 78 rpm disc, but the changer designed by RCA engineers called for a record with a 1-inch center cut-out.

Most historians have assumed that Victor's choice of a small disc implied that the firm's target was the popular music market. Nearly all popular records at that time were sold on 10-inch 78 rpm discs. But recent research by historian Alex Magoun shows that RCA reviewed its *classical* catalog and determined that nearly all its existing recordings had a duration of 5 minutes or less. With a 5-minute playing time, both the classical records and the shorter popular records could be sold on one type of disc. Clearly, engineers at Victor had different ideas about what constituted a classical song, since their colleagues at CBS had determined that a record for classical music had to play at least 20 minutes.

WAR OF THE SPEEDS

By the time 45 rpm record players went into production in the autumn of 1948, the LP had already been announced. RCA's dynamic leader, David Sarnoff, presented the 45 to the public as a major step in the perfection of the phonograph. The Victor catalog was revised to include scores of titles on the new discs, including albums (multirecord collections packaged in a box or book with sleeves for each record). The changer, first sold in March 1949, retailed for $39.94, which was about the same price as the Columbia LP player, and records cost 95 cents for Red Seals and 65 cents for most others.

The competition that ensued was dubbed by the press the "War of the Speeds," but in retrospect it is easy to see that both formats won. Record player manufacturers were easily able to accommodate both formats in addition to 78 rpm discs by offering multispeed players. The center-hole problem was easily corrected, although it demanded a spindle adapter to play 45s and the abandonment of RCA's first player design, which only played the 7-inch records with the large center hole. RCA began supplying the adapters almost immediately that snapped into the 45's center hole, but record players usually came with an adaptor to convert the center spindle. Despite RCA's efforts, the LP immediately captured much of the classical and album market, while the market for 45 rpm albums nearly disappeared. In fact, the word "album" in common usage came to be equated with the LP record. In subsequent years, the 45 would be used almost exclusively for singles. Unlike the battle between cylinders and discs in earlier decades, the War of the Speeds resulted in a compromise.

Both types of disc were also immediately adaptable to stereophonic recording, which had first appeared on tapes in the early 1950s but which was reintroduced by RCA for the LP and 45 discs in 1958 and later adopted by nearly every other manufacturer. Stereo required a record-cutting process that incised two separate recordings into a single groove, one on each of the V-shaped groove's walls. Playing stereo records demanded a new pickup, and a second amplifier and loudspeaker, making the conversion expensive and slowing stereo's adoption. Most records were released in both stereophonic and "monophonic" (as the old style of recording was now known) versions for many more years.

The 78 rpm disc did not disappear immediately, but lasted another decade as sales gradually dropped away. Perhaps the last 78 rpm disc ever pressed by a major American record company was Chuck Berry's "Too Pooped to Pop" in 1960, although there are documented releases of various songs on 78 rpm disc by a Finnish label in 1961 and there were probably many others later. In 1962, European record companies including EMI removed references to the remaining 78 rpm discs from their catalogs. Like most obsolete technologies, however, the 78 rpm recording refused to disappear completely. The band Moby Grape included a song on their 1968 LP entitled *Wow* that had to be played at 78 rpm to be heard properly. Many people retained the older records, as did archives and libraries. It was still possible after 2000 to purchase a new pickup and stylus to play the old discs, although nearly impossible to find a new record player that operated at 78 rpm.

By about 1960, high fidelity had been transformed from a specialized branch of electrical engineering to a part of mass culture. It was now a simple matter for consumers to purchase the technology off the shelves of their local retail store, rather than building it themselves. The term "hi-fi" came into currency as it became synonymous with the assemblage of electronic devices, loudspeakers, and players that comprised a listening system. Yet the key to this mass-market success was the availability of high-fidelity recordings, without which there would have been much less incentive for consumers to purchase such elaborate systems. Interestingly, the tape recorders introduced in the postwar period had a minimal impact on the success of hi-fi as a consumer product. Only in the 1960s would tape recording come into its own.

13

Revolution in the Studio

FROM DISC TO TAPE

The way that recordings were created in studios had already undergone many changes before the advent of tape recorders. The early acoustic recordings had taken place in cramped rooms with simple equipment and required musicians to alter the way they played or sang to suit the limitations of the recorder. Some of the limitations were removed in the 1920s with the advent of "electrical" recording technology. Larger studios, designed with careful attention to their acoustical properties, replaced the cramped little rooms. The new microphones could pick up sound well at distances from a few inches to several yards or more, allowing musicians to assemble in a more comfortable configuration. The use of electrical recording also helped force out the old traditions of the acoustic "recordists," which were replaced by the new techniques of "recording engineers" who focused on capturing every nuance of sound in high fidelity. Through all of this, however, it remained necessary for performers to deliver a perfect or near-perfect song in the studio, without the possibility of later editing. This remained the pattern through the end of the 1940s.

CHANGING PRACTICES

The mere appearance of the tape recorder in the record studios of the late 1940s did not in itself force changes on the recording process. In fact, tape recorders had been used to make musical recordings in Europe for over fifteen years beginning in the 1930s, with the most noticeable outcome being simply that the new machines could make somewhat longer recordings than 16-inch transcription discs. For reasons that have never been clear, the more creative uses of tape for recording music did not begin until the tape recorder was wrested from its legitimate corporate and institutional sponsors, who were mainly in Germany, and distributed around the world to new owners. These new owners included radio and television stations, motion picture studios, and record companies. While some recording engineers in these organizations saw the tape recorder as an instrument for capturing high-fidelity sound, others began to move beyond that and experiment with tape's other features, such as its ability to be easily edited. Like motion picture film, tape could easily be cut with scissors or a razor blade, removing parts or rearranging them, and then reassembling them using glue or adhesive tape. Recording tape could, with some difficulty, also capture two or more recordings, one on top of the other, the result being a combination or mixture. This technique was called "overdubbing."

Yet tape was not necessary to make high-fidelity recordings, and the perceived need for editing or overdubbing had not become so strong as to compel the introduction of optical recorders in the record industry. The state of the recording art had advanced to the point in the 1940s that engineers were fully confident in their ability to record music faithfully, with high fidelity. The introduction of the new high-fidelity media of the late 1940s, the LP and 45 rpm disc, did not depend on the tape recorders that were by then available. However, there were trends already underway in the making of records that involved manipulating the sound rather than simply capturing it truthfully, and these practices prepared the way for even more radical manipulations later, using tape. Some of the early examples of this included adding reverberation to recordings that would not otherwise have gotten it naturally from the studios in which they were recorded. Many studios of the 1930s and 1940s were constructed with sound-deadening materials on the walls, floors, and ceilings to absorb sound and prevent it from bouncing around the room. Eliminating a room's natural reverberation removed one source of uncertainty for the engineer, since the microphone sometimes "heard" reverberant sounds differently than the ear. What was recorded was not necessarily the same as what was heard, and so it was generally better to eliminate the reverberation that could overwhelm the music on a recording.

Removing the reverberation had the effect of "closing in" the performance, making it sound less like a live performance in a concert hall. That hardly mattered when the noise and distortion of consumer records and players masked the finer details of the music, but as the technology improved in the 1940s, engineers began looking for ways to reintroduce reverberation to give records a more "natural" sound. Some turned to making recordings in large performance halls rather than purpose-built studios, but that was not practical for every type of recording. Engineers also turned to the use of echo chambers and other technologies, resulting in a recording that was not in reality more "natural," even though the final results were often more pleasing to the ear. Echo chambers were rarely permanent installations, but often consisted of a loudspeaker and microphone set up at opposite ends of a long, reverberant space. Stairwells and restrooms, with their hard walls, were favorite places to create reverberation. Typically, reverberation would be created as a part of the recording process. Engineers tapped into the amplifiers in the studio's control room and transferred some of the signal down to the loudspeaker in the echo chamber, which broadcast the sound into the room. The same sound, now fortified with reverberation from the surfaces of the chamber, was captured by a microphone and fed back to the control room. All this had to be accomplished instantly, because of course it was all simultaneously being recorded onto a disc. Such manipulation of sound, rather than preserving the qualities of a performance as it was heard in the studio, purposefully distorted the performance in order to make the final product more appealing. It was just a little lie. When tape recording came to the studio, such methods of manipulation and distortion would blossom.

EDITING

Until the late 1940s, recording engineers making phonograph records abided by the strict limitations of direct-to-disc recording. In the motion picture industry, recording engineers depended on the ability to edit the recordings they made, removing unusable parts of them, rearranging bits, adding background music to dialog, and so on. There was no editing or rerecording possible in the making of a phonograph record except under extraordinary circumstances. Performers practiced their pieces until they got them right, then went into the studio to record. As a result of the process used in the record industry, recording sessions were brief. A performer might spend a few hours in the studio and make one record every 20–30 minutes. If a musician missed a note, the record company would decide if it was noticeable enough to warrant making another recording. The expense of doing so made it imperative to get it right the first time, yet apparently the expense and inconvenience were not so great that the record companies seriously considered using optical sound recorders.

With the arrival of tape, that pattern began to change, as did the respective roles of musicians and recording engineers. The downward direction of the price of master recordings, which began in the late 1930s with the introduction of the acetate disc, was pushed further with tape, which was even less expensive to use and which could, if necessary, be reused any number of times by erasing it. Tapes could be easily trimmed or edited, so that it was no longer necessary to begin and end a recording with precise timing. If the recorder was switched on a few seconds early, that part of the tape could be cut off and discarded. Two pieces of tape with different recordings on them could also be joined by splicing without any audible transition or degradation of the final product. In this way, tapes of short songs could easily be assembled into an "album" to be recorded on an LP. Tape editing became common for nonmusical recordings such as broadcast news, where long interviews could be edited down to their essentials. Music recordings could also occasionally be edited to replace flubbed sections, although this was a difficult process. Yet the adoption of editing as a technique in recording occurred quickly, and was already a common procedure when the first splicing block, a tool for making clean splices in audio tape, was introduced in 1949. The capacity to edit and the lower cost of recording reduced the risks of not capturing the recording the first time, and these two factors probably encouraged even greater experimentation with tape in the studio.

As getting a performance right the first time became less of an imperative by the late 1950s, there was a relaxation of expectations for performers in the studio, and this too would come to have an influence on how recordings were made on tape. Established musicians, bands, and orchestras seem to have carried on much as before in the 1950s, arriving at the studio with polished pieces ready to perform, or having the skill necessary to read the songs from sheet music with perhaps just one practice playing before recording began. But the new generation of recording artists would not always be so well prepared. With the popularity of rock and roll music came performers who were less accomplished musically, or who had less formal training. By the 1960s, musicians would often show up at recording studios with no rehearsed performance at all. The tape would roll on take after take until an acceptable performance was captured. The cheapness of tape helped stretch the time of the average recording session over the course of several decades from a maximum of a few hours to days, weeks, or even months.

Simultaneously, engineers and musicians were beginning to use tape experimentally in new ways. With tape, it was possible to make a recording in two or more parts, mixing them together later to make the final recording. This was being done even in 1950, using a tape recorder, for example, to capture the instrumental parts of songs, and then "overdubbing" the vocals onto the instrumental part later. Other practices emerged, such as the making of novelty records made by mixing recordings made at one speed with recordings made at another. This technique was immortalized by Ross Bagdasarian in the 1950s with his "Chipmunks" records. To make these discs, he accompanied himself on an overdubbed tape singing the parts of Alvin, Theodore, and Simon Chipmunk, rerecorded them at high speed to give the voices a high pitch, then added additional layers of instruments and voice at normal speed to complete the recording.

STEREO: REALISM OR DISTORTION?

Progress in designing technology for making recordings stereophonically (two or more channels, recorded and played simultaneously) was well underway by 1950, even though few consumers had yet heard of it. The introduction of tape recording would propel stereo toward commercial reality. Yet it remained a question whether stereophonic sound was another step toward high fidelity, or whether it was a form of intentional distortion of musical performance. Bell Telephone Laboratories had made special, two-channel recordings on disc as early as 1932, but they were not released to the public until the 1960s, when they were rerecorded and pressed on LP records. The original LP and 45 rpm records were all single channel and remained that way well into the 1950s.

Yet recording engineers enthusiastically pursued stereophonic sound in the early 1950s, inspired by the simplicity of converting single-channel tape recorders into two- or three-channel stereophonic devices. Converting to two-channel tape, for example, required only that the normal recording head, which "wrote" the recording across nearly the entire surface of the tape, be replaced by two half-width recording heads, each of which wrote on just half the tape. Consumer tape recorders using a single, "half-track" head were already in production; they made it possible to "double" the recording capacity of a reel of tape by recording along half its width, then flipping it over to record on the other side. Two-channel tape recorders using two half-track heads were demonstrated by the Magnecord Corporation around 1949, although the Magnecord line was priced well above most consumer tape recorders, limiting its potential market to the professional market. Stereophonic, open-reel tapes and consumer recorders appeared in 1954, produced by a tiny U.S. company called Livingston Audio Products. A year later, the Voice of Music Corporation, then a major consumer electronics manufacturer, began to offer an inexpensive kit to convert its existing line of recorders to stereo recording and playback. A few small companies then began to offer consumers stereophonic tapes for these machines, but it was not until 1957 that a major record company, Capitol Records, issued such tapes. Tape would not emerge as the mass-market form of stereo listening for many years, however. Even the entrance of the mighty RCA-Victor into the stereo tape field in 1958 was not enough to entice consumers. Tape recorders and tapes were still too difficult to use and much more expensive than phonographs. In fact, all forms of tape recording were overshadowed commercially by the phonograph until the later 1960s.

Initially, the stereophonic phonograph would not be easy to sell to consumers, either. The only stereophonic system that reached the market before 1958 was invented by Emory Cook, an audio enthusiast and talented engineer. Cook in the early 1950s became interested in "binaural" recordings, where headphones are used to separate the channels of a two-channel recording. When properly made, binaural recordings give the listener a realistic sense of the source of sounds. Binaural recordings are sometimes made today, but they are much less common than stereophonic recordings, which are usually defined as two-channel recordings intended to be heard with both ears via loudspeakers rather than headphones. Cook in 1951 demonstrated a type of phonograph record with two concentric grooves, similar to the system developed at Bell Labs years earlier. He subsequently manufactured and sold such records along with a special, dual-pickup tonearm used to play them.

While the Cook system would see only minor commercial success, another technology inspired by the Bell Labs experiments would come to light by 1957. In that year, the Westrex Corporation (successor to ERPI, the distributor of Western Electric motion picture equipment) began demonstrating what it called its 45-45 stereo disc recorder. Using essentially the same groove dimensions as the LP or 45 rpm disc, the 45-45 cutter incised a recording into each wall of the V-shaped groove. The name of the machine came from the fact that the grooves were set at 45 degrees from horizontal. At nearly the some moment, Decca in England was readying a system similar to that developed by Blumlein in the 1930s, and which was quite similar to the Westrex recorder. By common agreement, record and player manufacturers adopted the 45-45 system late in 1957, and in 1958 set industry standards for the details of the groove, stylus dimensions, and other technical specifications. While they may have hoped to make the new discs playable on existing sets, in the end they were forced to adopt a smaller stylus tip for the stereo records. Consumers who wanted to hear stereo would have to purchase either a whole new player, or a new pickup and second amplifier and loudspeaker. RCA and others introduced their first stereo records late in 1958, along with inexpensive record players fitted with a flip-over stylus for both ordinary and stereophonic records.

Many of the earliest stereophonic records were made to demonstrate the stereophonic effect as dramatically as possible. Recordings were made of things in motion—people talking, trains, and so on—and when reproduced on a stereophonic system, listeners could clearly hear how the sound seemed to pass from one speaker to another. The real intent of stereo, however, was to add a sense of three-dimensionality to recordings of music. Consumers were told that stereophonic recordings sounded better because the sound was more "realistic." That is, stereo was marketed to consumers as an innovation in high fidelity, another step toward a system that could reproduce in the home the sound of the concert hall.

Music critics and recording engineers were not convinced. Controlled stereo-listening tests were conducted in which a live performance was picked up by pairs of microphones and transferred in stereo to listeners in another room. Listeners could not accurately describe the precise location on stage of the original performers, although they believed they could. By varying the placement of the microphones, it was possible to, in effect, shift the location of the performers as perceived by the listeners. Music critics, relying solely on their own ears, argued that in a real concert hall, it was difficult to locate by ear the musicians in a band or orchestra, while it was possible with a stereophonic recording. However, they agreed that the stereophonic illusion was fascinating to hear, and that it did not detract from the enjoyment of the music. In other words, while most listeners found the stereophonic effect pleasing, it was rarely able to accurately re-create the original placement of performers. It was, therefore, more of a distortion of the original sound than a step toward higher fidelity. The point is subtle but important.

As rock music gained popularity in the 1960s, studios began using the stereo effect to create larger distortions of this type, usually intentionally. In a day when "singles" on 45 rpm discs still dominated the rock and roll market, making a song sound good on a 45, which usually meant making it sound good when played on an inexpensive record player, was more important than achieving stereo realism or catering to the high-fidelity enthusiasts.

Often, two (or more) versions of a song were made, the first in one-channel (by then called "monophonic") sound for the single, and a second, stereo version created later for the LP. Stereo effects on LP were often included simply as a gimmick to attract customers to the higher priced album. Typically, stereo recordings of pop and rock singles were remixed versions of the monophonic master tapes, so that the electric guitar track was put on one channel, the bass guitar on another, and the drums and vocals fed equally to both to make them seem to come from the center. They were not necessarily originally recorded in stereo. In other cases, recording engineers "panned" the sound of an instrument or voice from one speaker to another and back again, hoping to convey the "psychedelic" sound that had become popular late in the 1960s. Such effects were clearly not related to high-fidelity sound, although they were achieved using hi-fi technology.

MULTICHANNEL RECORDERS AS CREATIVE TOOLS

Music could be heavily manipulated in other ways through the use of multiple-channel recording in the studio. As early as 1956, popular musician Les Paul had installed in his home studio an Ampex tape recorder capable of recording eight parallel tracks on a 2-inch-wide tape. The recorder circuitry allowed a recorded track to be played while a recording was being made on another track, a feature that Ampex called Sel-Sync. Paul, who had been experimenting with dubbing techniques since the 1940s, now used the multitrack machines to record his songs in separate pieces, mixing different tracks together to produce new combinations until he was finally satisfied with the result. While Paul was neither the inventor of multitrack recording nor the first to experiment with these techniques, his prominence brought multitrack recording to the attention of recording engineers and musicians.

TRACK "BOUNCING" AND TAPE

In the early and mid-1960s, multitrack studio recorders used two tracks on a standard quarter-inch-wide tape or three tracks on a half-inch-wide tape. The purpose of using multiple tracks was not necessarily for recording in stereo, but rather it was useful for mixing and overdubbing. Many songs of that era were made in stages by exploiting multitrack and Sel-Synch capabilities. Typically, the drums and rhythm would be recorded on one or two of the tracks, and when it sounded right, they were mixed together and fed to the third track. This freed up the other two tracks for more recording. Then the background and lead vocals and any other necessary instrumental parts were recorded in the same way before being "bounced" down to the final mix used to make the record. Often, studios would leave the vocals and music on different tracks. As soon as the instrumental track sounded right, the musicians could be sent home and the vocals worked on later, as was frequently necessary. This saved the studios money. Four-track recorders became a standard type of machine by the mid-1960s for this kind of work and remained in use for many more years. It was also possible, though more difficult, to use two or more multitrack recorders in this way, although it was a challenge to synchronize them perfectly.

Increasingly, the multitrack tape recorder and the recording engineer or mixer became part of the artistic process. A landmark in the creative use of tape in studios was the 1966 Beach Boys album, *Pet Sounds*. It took Brian Wilson, the group's lead songwriter, three months, five studios, and seventeen separate recording sessions just to finish one of the songs, "Good Vibrations." Like some of Wilson's earlier works, the song was a densely layered, nearly symphonic approach to pop music, and though many considered the Beach Boys a "bubblegum" group, other musicians took note of their studio accomplishments. Although he used up-to-date 3- or 4-track recorders, Wilson preferred to mix songs monophonically; stereophonic versions were created later for the album releases, or not at all.

Even more dramatic demonstrations of the possibilities of multitrack tape and stereophonic reproduction were the late-1960s recordings of the Beatles. Their mentors at Abbey Road studios in England developed some of the same complex layering techniques and studio gimmicks as Wilson to record the acclaimed *Sgt. Pepper's Lonely Hearts Club Band* album over 129 days in a series of sessions in 1966 and 1967, using two separate, 4-track recorders. The songs on *Sgt. Pepper's* conveyed the group's new fascination with drug culture and Eastern mysticism through the clever use of what could be best described as novelty effects—bits of tape played backward, echo and reverberation, intentional distortion, and the gimmicky (from today's perspective) use of stereophonic reproduction. Like *Pet Sounds*, the album was notable for the amount of time that the artists spent in the studio, as well as the enormous labor that went into the mixing process. By one account, the album took 700 hours to record and mix.

Through all of these changes, the ultimate product of the studio remained the phonograph disc. After recording, editing, and remixing, the final "master tape" was sent to another part of the studio where it was used to make a disc recording. Then this disc recording would be transformed into a mother disc, master disc, and stamper in the usual way. Through the end of the phonograph's lifetime, high-quality studio disc recorders and recording techniques remained very much in use, though the focus of the recording process was now the tape recorder.

IN THE AFTERMATH OF TAPE

Recordings of the late 1960s that heavily exploited the possibilities of multitrack tape helped create a demand for studio tape recorders with even more tracks. The first 8-track studio machines became available late in 1967, and they were followed by 16- and 24-track recorders by 1970. Along with this new technology came new expectations about recordings among both musicians and listeners. Musicians, who were learning more about the recording process and beginning to take a more active role in it, came to rely as much on studio technology to shape the final product as on their own musical talents. Bands no longer saw it as necessary or even desirable to come to the studio prepared to give a perfect performance as a group. It could all be fixed "in the mix," although that often involved monumental labor on the part of the recording engineer. Further, the role of recording studio personnel expanded and became part of the creative process, as the technology began to be used more intensively to shape sound.

The use of tape recorders reshaped sound in many ways. At first the quest for musical perfection led almost inevitably toward an ever-greater interest in editing as a way to correct performance errors. But at the same time that editing improved a recording, it also took it farther from its origins as a live performance. As editing and dubbing became less about "fixing" and more about creating something new the definition of high fidelity had to be adjusted in the posttape world to move beyond the old aim of simply simulating a concert hall performance. To the listener, this shift in the philosophy behind recording may not have been all that noticeable. Clearly, though, consumers were satisfied with the results. Sales of records skyrocketed in the late 1950s, and again in the late 1960s, because along with changes in musical tastes, shifting demographics, and a healthy economy, the technology of high fidelity helped sell records. The use of multitrack recorders transformed the recording process. It became routine for performers recording a song to make not one recording of a complete song but many partial recordings. The resulting recordings were not simply improved versions of studio performances; they were constructed or synthesized from a collection of components, assembled from pieces like any other complex technological product. As such, they could not usually be performed outside the studio, such as when a band played live before an audience. By the 1960s, especially in rock and roll, performers were creating records that were so far removed from what was possible to perform live that onstage renditions of recordings were sometimes barely recognizable to audiences. Tape had become something of a musical instrument, but one that was difficult or impossible to "play" anywhere else but the studio.

14

Mobile Sound

THE PREHISTORY OF PORTABLE SOUND

Mobile listening has become an important part of aural culture in the last fifty years, since the introduction of the transistor radio. Many people possess technologies that let them listen to recordings as they travel on foot, as they drive in their cars, as they fly in airplanes, or nearly anywhere else. People have taken music to the bottom of the sea and to outer space (the farthest a recording has gone as of 2004 is about 13 billion kilometers, on a disc mounted on the *Voyager* deep space probe). Today people in much of the world can take sound recording technologies with them nearly anywhere they go. Yet the roots of mobile listening stretch back to the nineteenth century.

Early phonographs and graphophones, for example, were relatively portable. Many of the machines of the 1890s and later came in small, enclosed cabinets with a carrying handle. Because most early record players used spring-driven motors rather than electricity, there were no batteries or cords to deal with. A big boost in the sale of portable record players occurred in the post–World War I period, when sales of automobiles were taking off. Inexpensive, lightweight record players were marketed in the United States and Europe for those who wanted to bring their music with them on the picnics and outings made possible by the auto. Yet these devices could not be played inside the automobile as it was moving. Playback of a record was too delicate an operation to be undertaken in a bouncy automobile.

AUTOMOBILE SOUND

True in-car entertainment had to await the introduction of the automobile radio. Radios specifically designed for the automobile appeared in the late 1920s, and the radio became one of the most popular, but also one of the most expensive, items of optional equipment for cars during the 1930s (often 10 percent of the purchase price of the auto itself). Unlike today's car audio systems, which stuff everything into a single enclosure, early car receivers consisted of several subassemblies: a tuner that was typically mounted on the steering column, a second enclosure containing the electronic amplifiers and circuits, and a box containing the special high-voltage power supply required by vacuum tubes.

Advertisements of the day suggest that manufacturers saw the appeal of the radio in the car as a medium for musical entertainment. That may not seem surprising today, but it must be remembered that in the 1920s and 1930s much of radio programming was nonmusical. It was more like today's television, with comedies, dramas, and other theater-like programs. Manufacturers correctly perceived that consumers wanted to use music for its mood-altering potential, calming them in heavy traffic, for example, or helping to excite them when driving for fun.

Sales of auto radios grew almost without interruption even during the Depression years, when overall auto sales were rapidly declining. Given that the radio was both expensive and optional, and that buyers could even—but rarely did—request to "delete" the radio from a car they intended to buy in order to save money, it is evident that the radio had a powerful appeal to consumers.

MOBILE RECORD MAKING

As long as the phonograph was the only recording medium generally available to consumers, it was unlikely to become a competitor to the radio receiver for in-car entertainment. There was, however, a certain small segment of the market where an in-car phonograph seemed to make sense: office dictation. Makers of office dictation equipment, from the late nineteenth century to the present time, have tried to convince

Automotive sound systems in the form of car radios became popular in the early 1930s. There were even a few proposals for in-car recorders. Here, a skeptical Edison looks on as an automobile dictation system is demonstrated. U.S. Department of the Interior, National Park Service, Edison National Historic Site, West Orange, New Jersey.

potential consumers of the benefits of mobile dictation. The Dictaphone Corporation, for example, sponsored the writing of Sir Edward Lund's book *Round the World with a Dictaphone* in 1926. The book had little to do with the Dictaphone, but the preface mentioned that the author had used it to capture impressions of his round-the-world trip, and later used the recordings to write the book. Company officials could point to this book as a model of what could be done with the business phonograph outside the office. Thomas Edison promoted much the same idea, having his picture taken for the press in an open car, using one of his firm's dictation machines to capture a memorandum.

When magnetic recorders became available after 1945, automobile dictation enthusiasts finally had a device that could reliably record and reproduce sound despite the vibration present in a moving automobile. Several inventors took existing wire or tape recorders, modified them for use with the automobile's 12-volt electrical system, and demonstrated their usefulness for in-car dictation. Dictation machines mounted inside automobiles have appeared and reappeared every so often ever since. The Austrian electrical firm Stuzzi marketed a dictation machine built especially for dashboard installation in the early 1970s, and General Motors (among others) offered optional combination radio-cassette recorder/players in the early 1980s. None of these devices saw widespread use.

HIGHWAY HI-FI

Neither would Highway Hi-Fi, the phonograph's second-to-last appearance in the automobile, have much success. Peter Goldmark at CBS Records had pushed this technology forward, and it made its way to the dashboards of Chrysler cars between 1955 and 1960. Using a 7-inch disc turning at just 16⅔ rpm (approximately half the speed of the LP), with a groove that started near the center of the record and ended at the outside edge, Highway Hi-Fi was an ambitious if ultimately unpopular Chrysler option. With a tonearm that pressed so hard on the records that it wore them out quickly, Highway Hi-Fi could play a record without skipping, but in the end it was too expensive, and too few records were available. What was, most likely, the last of the automobile phonographs were offered in the 1960s by Phillips, a European manufacturer, and by RCA. Both the RCA player and the Phillips "Auto Mignon" used ordinary 45 rpm discs. Like Highway Hi-Fi, the spring-loaded tonearms on these players attempted to eliminate skipping by simply pressing the needle tightly into the groove. These attempts to offer consumers a way to play their own music in the automobile were technically flawed, but manufacturers had identified an unfulfilled consumer desire. Satisfying it would have to await a new technology.

THE ARRIVAL OF TAPE

Tape recording technology had a proven ability to perform well under less than ideal conditions. It was not long before tape recorders were experimentally installed in automobiles, and then offered as a regular production item. Tape was one of the two technical innovations that heralded a new age of in-car entertainment beginning in the late 1950s. The other key innovation was called the transistor. This electronic component first appeared in 1947, invented at the Bell Telephone Laboratories. The transistor is today the basis of virtually every electronic system, from radios and televisions to computers. It was a replacement for the older technology of the vacuum tube, and was able to amplify signals while taking up much less space. Further, it required less power and used lower voltages, making it better suited for use in battery-operated systems. A transistor radio, for example, required only a single, small battery in order to operate. It could perform well with batteries of 9 volts or less, unlike most vacuum tubes that demanded 50 volts or more to work well. In fact, earlier vacuum-tube car radios required an expensive, unreliable electromechanical device to derive high voltage from the auto's 6- or 12-volt system. The transistor's early consumer applications included handheld, battery-operated radios, which first appeared in late 1954 and which became a mass-market item the next year. Tubes were used for several more years in car radios, but were replaced by transistors during the early 1960s. A small, battery-powered transistor tape recorder, called the Mohawk Midgetape, appeared in 1955, but it would be several more years before battery-powered tape portables would catch on. Tape recorders required additional current to supply their electric motors and recording circuits, so practical battery-operated models awaited improvements in efficiency. Despite these engineering obstacles, it was clear that tape was better suited for mobile entertainment than the phonograph.

PORTABLE TAPE

The transistorized tape player found its first large consumer market among automobile owners. The early automotive tape players, introduced in the middle 1950s, had a rather unusual birth and early childhood. They were inspired by a technology to show motion pictures. In the 1950s, the "low end" of the movie business consisted not of Hollywood blockbusters or even made-for-television features, but of advertisements, educational films, and even pornography. These films were distinguishable not only by their content but also by the technologies used to produce and exhibit them. The endless-loop projector, which could be used to show films repetitively, was widely in use from the 1930s onward to show short movies such as advertisements or announcements in exhibit halls. One form of the endless loop projector, made by an American firm called Television Associates, used a loop of 8mm film enclosed in a small plastic cartridge.

It was this cartridge that may have inspired an otherwise obscure inventor from Ohio to create a new form of tape playback device. Substituting ¼-inch magnetic tape for the 8mm film, Bernard Cousino created the Audio Vendor, a tape cartridge for continuous playback. Cousino in 1953 began marketing the Audio Vendor, which could be played on any reel-to-reel tape recorder, as a way to create audio-enhanced, point-of-sale displays for products in stores.

George Eash, one of Cousino's associates, later patented an endless-loop tape cartridge requiring a player designed especially for that task. In Eash's design, tape cartridges came in several different sizes, holding anywhere from about 10 minutes of tape up to several hours' worth. Instead of manufacturing the recorders and cartridges himself, he instead licensed the manufacturing rights to other companies, beginning with Telepro Industries of Cherry Hill, New Jersey. Telepro marketed its Fidelipac system as a way to provide background music in stores or restaurants, serving a market where there was a growing demand for inexpensive background music systems. Once started, the machine required no rewinding or changing of tapes, and after the musical program ended, the machine could restart it immediately.

Soon other licensees also began offering similar products, including the Viking Corporation of Minneapolis, a relatively small maker of tape recorders and other products. Viking encouraged a local radio station to begin using the tape cartridges to hold commercial announcements and other messages. Because these messages were played often, on ordinary discs they wore out quickly. Tape is more durable, but this advantage was outweighed by the bother of threading up a tape in an ordinary reel-to-reel recorder. The tape cartridge could easily be inserted and removed, and by setting a switch, the Viking tape player would automatically detect the beginning/end of the loop and stop the tape, preparing the cartridge for its next playing. Radio stations around the world adopted the "cart" in the 1960s, and new firms such as Audiopak and International Tapetronics Corporation began to specialize in the manufacture and sale of cartridges, players, and player-recorders. This technology was in use through the early 1990s when it began to be replaced by compact discs and digital recordings.

MADMAN MUNTZ

The Fidelipac cartridge also came to the attention of entrepreneur Earl Muntz of California. Muntz began his career as a car salesman, turning to radio manufacturing in the 1950s and then selling his own line of inexpensive television sets, custom manufactured for him and employing a notoriously cheap design. In the early 1950s, he had briefly become a manufacturer of an exotic, custom-made car called the Muntz Jet, but sold only about 400 of them before giving up. He advertised himself as

"Madman" Muntz; his wife, he claimed, thought he was mad for selling his TV sets so inexpensively. By the late 1950s, Muntz Television was ailing, and Madman Muntz ceased TV production in 1959. Looking for a new product, he turned to the idea of an automotive tape player. By using the smaller of the Eash cartridges, the same size then in use in radio stations, and having an inexpensive player custom manufactured in Japan, Muntz created an exciting new automotive accessory. He found that the music industry was willing to license his company to duplicate top-selling albums on Fidelipac cartridges. Stereophonic recordings were by then widely available, so Muntz adopted this still-new technology as a standard feature of the players (although they would play monophonic tapes as well). The Muntz players were designed to play a full stereo album without flipping the tape or changing channels. The reasonably high-quality players could even boast some safety features, such as the minimization of the number of knobs and controls, which made it easy for the driver to concentrate on driving.

The whole system was marketed beginning in 1962 under the trade name Muntz Stereo-Pak. The transistorized players, at $80–$170 (later $49.95), cost more than the average automobile radio but offered consumers something different. Tapes, at $3.50–$6.00, were also somewhat more expensive than comparable LP records. However, the public enthusiastically embraced the new technology. Muntz added franchised dealers in Texas and Florida, and sold the players from his own stores around Los Angeles. In April 1962, the booming business of making tapes led him to cancel his arrangements with local tape-duplicating firms and begin producing his own tapes at a new facility in Hutchinson, California.

The appeal of the Stereo-Pak lay in its capacity to allow consumers greater control over the music they listened to in the car. Unlike radio broadcasts, the Stereo-Pak never delivered advertisements or other unwanted annoyances. The system also had a strong emotional appeal to those who craved the latest technical gadgets. Further, in the context of California's famous car culture, a technologically advanced accessory was likely to be warmly received. Muntz had an instant hit. Ever the showman, he made sure that Hollywood stars installed Stereo-Pak players in their cars, and his press releases in 1963 claimed that Muntz players were already in Frank Sinatra's Buick Riviera, Peter Lawford's Ghia, James Garner's Jaguar, Red Skelton's Rolls Royce, Lawrence Welk's Dodge convertible, and even the automobile of dashing conservative Senator Barry Goldwater of Arizona. During 1964 and 1965, several record companies also began issuing releases on the new format, and the Stereo-Pak began to attract national attention.

FROM STEREO-PAK TO STEREO EIGHT

One of the people who took note of the success of the Stereo-Pak was William P. Lear, entrepreneur and leader of the Learjet aircraft company. Lear Radio, the company he had owned in the 1940s, had manufactured wire recorders under an Armour Research Foundation license, and Lear had personal ties to the Motorola Corporation, then a major manufacturer of auto radios. In 1963, Learjet became a distributor for Muntz Stereo-Pak, with Lear intending to use the system to provide soothing background music in his forthcoming Learjet business aircraft. However, he later decided to reengineer the player and manufacture it himself. He designed a slightly simplified cartridge that was cheaper to manufacture and that bypassed some of the earlier Eash and Cousino patents. He also asked the Nortronics Corporation, a maker of tape recorder heads, to make a head capable of playing a tape with eight parallel tracks. The Muntz system used four parallel tracks, representing two programs (or album "sides") of two stereophonic tracks each. Two programs times two channels equaled four tracks. Lear's system would squeeze twice as much music onto the tape, using eight tracks, representing four stereophonic programs of two tracks each. He called it the Stereo 8, but it is more commonly remembered as the 8-track.

In 1964, Learjet constructed 100 demonstration players and distributed some of them to executives in the automobile industry. The Ford Motor Company soon agreed to offer the player as optional equipment on some of its 1965 models, with Motorola (already a supplier for Ford) to make the players. At about the same time, Lear successfully negotiated a deal with RCA Victor to permit him to create a library of 175 recordings for initial distribution.

1965 sales of Stereo 8 systems surprised executives at Ford, who made sure that more models were offered with players the following year. In 1967, already about 2.4 million of the players were in use. Learjet made access to the patented design relatively easy to acquire, and many other manufacturers joined Motorola. By 1968, General Motors and Chrysler were also offering players. Perhaps more importantly, several firms including Learjet were offering home 8-track players, betting that consumers would appreciate the ability to buy tapes that could be played in the car or in the living room. It was a winning combination. Compared to the lackluster sales of reel-to-reel tape recorders after they were introduced in the 1950s, the 8-track was a spectacular success. By the early 1970s, it was becoming a common feature of new autos, and owners were retrofitting older cars with players offered by a multitude of aftermarket manufacturers. By the middle

The flowering of in-car tape systems came after the introduction of the Stereo 8 (or 8-track) system in 1965. CORBIS.

of the decade, the format had captured about a third of the record market in the United States, and had spread to several countries in Europe. Yet within five years, record companies would begin dropping the format. The reason: a competing type of tape cartridge called the Compact Cassette.

THE CASSETTE'S HOLD ON MOBILE LISTENING

The Compact Cassette was introduced in the United States in 1964, a year before the 8-track, and was a little-noticed entrant among a group of tape cartridge formats that appeared about the same time, all of them now forgotten. The cassette consisted of a narrow tape about half the width of the standard quarter-inch tape used in the Stereo 8 and nearly every other tape application. Its corporate sponsor was the Phillips Electronics Company of the Netherlands, a major European consumer electronics firm that sold its products in the United States under the brand Norelco. The cassette's intended market was ill defined in 1962 when it was first introduced in Europe. The small, battery-operated cassette recorder, called the Carry-Corder, cost a hefty $79.95 and blank tapes were about $2.00. Recorded cassettes would not be available for years. The machine's performance was markedly inferior to open-reel recorders costing less than half that much. It was not even equipped with the necessary mechanisms to make it useful as a dictation machine, although it was clearly not intended to be used for dictation since it would have competed with the company's existing line of cartridge-tape dictation recorders. Its primary selling point was not sound quality but, like the 8-track, convenience and portability. Within a few years, Phillips licensees and tape manufacturers had brought the price for recorders down to $29.95, and tapes cost as little as 75 cents. The Compact Cassette system began to catch on among a group of consumers who were already buying the cheapest open-reel machines offered by Japanese manufacturers: children.

Just as children constituted an exploding market for recorded music in the postwar period, so too were they an important, growing market for consumer electronics. When several Japanese firms began offering open-reel tape recorders in the early 1960s at prices under ten dollars, and these recorders began appearing in department stores, parents began buying them in greater numbers. By the early 1960s, when the Japanese had captured virtually the entire U.S. market for tape recorders of all types, the majority of recorders imported were battery-operated portables, and these imports were usually among the least expensive models on the market. One survey of children's listening habits identified adolescent girls as the primary users of such tape-recording equipment, although it is now clear how patterns of use developed later. Because of the general lack of availability and high cost of recorded tapes, children tended to make their own, either taping the output of their record players or recording broadcasts off the air. By the late 1960s, record companies had detected the trend and began offering limited selections from their catalogs on cassette, but the technology of cassette duplication had not kept pace and the sound quality of these early releases was decidedly inferior to that of records and Stereo 8 tapes.

This market was ideal for the Compact Cassette system, which emphasized simplicity, low cost, and ease of use, at the expense of high fidelity, stereo, the ability to edit, and many of the other features that had originally been thought of as tape's primary advantages. But hi-fi did not remain in the background for long. The cassette demonstrated its potential appeal to adult hi-fi customers when a few companies began offering home cassette "decks" in the late 1960s The term "deck" had come into use in the early 1960s when numerous manufacturers began selling open-reel tape recorders that did not include amplifiers or loudspeakers, but had to be connected to a separate hi-fi system. With such a tape deck, it was easy to transfer records or broadcasts to tape to create an inexpensive second copy, perhaps for mobile listening. While tape equipment for making such copies had been available since the late 1940s, copying had never been so simple and inexpensive, nor had there been a technology so well suited for portable tape listening.

HI-FI AND THE CASSETTE

When the first Dolby Corporation noise-reduction system for cassette tapes made its debut in 1970, the cassette truly began its transition from a toy to a serious high-fidelity instrument. The Dolby system used electronic circuits to alter the recording and playback characteristics of a cassette deck, compensating for the cassette's inherently high level of background "hiss" and lack of high frequencies. The Advent Corporation was apparently the first in 1970 to offer an add-on Dolby "black box" as an expensive accessory for its line of home cassette decks. Later that year, Advent, Fisher Radio, and the Harmon-Kardon companies began selling a high-fidelity Dolby cassette deck made by the Nakamichi Corporation of Japan (then a manufacturer of consumer electronics sold in the United States under various other brand names). Tape manufacturers responded by introducing a new type of tape made especially for cassette systems, which had chromium dioxide mixed in with the normal iron powder, resulting in magnetic properties that were better suited to the cassette systems' slow speed of just 1⅞ inches per second. These "chrome" tapes, introduced in 1971, along with numerous technical improvements aimed at overcoming the cassette's deficiencies, resulted in both a better sound and a better image for the technology, and helped transform it into an audio technology for adults.

The cassette's success was based on its capabilities as both a recorder and a player, so sales of both blank and recorded cassettes were important. The market for blank 8-track tapes (or, earlier, blank discs) did not amount to much until the late 1970s, even though 8-track recorders were available from the start. Sales of blank cassettes, on the other hand, far outstripped sales of all other types of blank tapes and soon became the basis of a major industry segment. Recorded cassette sales, which had been almost nonexistent in the 1960s, began to eat away at 8-track tape sales by 1975, and during the next five years met and then surpassed 8-track levels. Already in 1979 one journalist wrote an "Obituary for the Eight Track," calling it unnecessary in the era of the cassette.

The 8-track held on to less than half the total market for recorded tapes in 1979. The economy had been in a sorry state for several years prior to this, and the record industry was looking for ways to cut costs. In 1982, most major record companies ceased producing 8-track tapes, instead sending the business to independent tape-duplicating firms. Then, in 1983, they dropped the format altogether, except for a few top-selling new releases and the tapes made available to mail-order "record club" customers. Retailers, bothered by the necessity of stocking albums in a multitude of different formats, were all too happy to stop stocking the tapes.

PORTABILITY EVERYWHERE

What began around 1910 as a way to carry music along on trips became in the 1970s an ingrained pattern of consumer behavior. Patterns of mobile listening emerged as a mass phenomenon in the late 1950s, following the introduction of the battery-powered transistor radio. They matured in the 1960s with the introduction of both the Stereo 8 automotive sound system and inexpensive transistor tape recorders, each marketed toward a different group of consumers. Ultimately, the consumer populations attracted to these two technologies would merge in the 1970s and would be joined by others. Notably, in-home listeners would begin to use the cassette to make tapes, which they would then listen to on a variety of mobile players. Ultimately the cassette, with its great flexibility and improving sonic qualities, would steal the market from the 8-track in the early 1980s.

Sales statistics reflect the shifting tides of consumer behavior. United States sales of tape recorders, for example, which had hovered around a few hundred thousand in the late 1950s, totaled over 12 million units in 1975 alone. That massive increase came at the expense of phonographs of which only about 1.3 million were sold in 1975. That same year, over 162 million blank cassettes were sold in the United States alone, representing about 80 percent of the market for blank recording media (most of the rest representing sales of blank 8-tracks and open-reel tapes). Blank and recorded cassettes were also feeding a huge number of new automobile tape players. In 1977, the first year in which the federal government published detailed statistics, about 12 million automotive tape players were sold (Morton 1995, 518–519).

Underlying the cassette's success is a change in consumer behavior that foreshadowed changes to come. Enthusiasm for making recordings, in addition to listening to them, emerged in the 1960s among very young consumers for cheap tape recorders, but was gradually transferred to a broader segment of the public by the 1970s, perhaps as the children of the 1960s grew up. Home recording was not an end in itself for most consumers, but a way to create mobile versions of the music they liked. During these same years, consumers were moving beyond the level of portability permitted by transistor and auto radios, and embracing the habit of choosing their own music for mobile listening (although radio listening also remained very popular). The perceived need for choice and mobility that helped the cassette rise to commercial dominance would profoundly shape the development of virtually every subsequent audio technology as well.

15

Cassette to Compact Disc

THE RISE OF THE CASSETTE

In 1975, U.S. sales of LP records totaled over 257 million units. Behind the LP in sales with over 164 million units was the 45 rpm disc, followed by the 8-track tape cartridge with 94 million units. In last place was the cassette, with just 16 million units sold. Ten years later, the 8-track was obsolete, and five years after that, both the LP and the 45 were virtually extinct. Behind these changes was the rise of the Phillips compact cassette. Introduced in 1962, the cassette (as it was generically known) found early acceptance among the young, who used it to make copies of phonograph records or record music off the air. As engineers found ways to improve the sound quality of the cassette, adult consumers later adopted it, using it both in the home and in portable and automotive players. While the 8-track was the pioneering medium for mobile music, the cassette stepped in to offer consumers greater convenience at a lower cost.

HOME TAPING

One of the key aspects of the history of the cassette tape is the role that home taping played in popularizing the format. Detailed statistics for blank tape sales are not readily available, but the evidence suggests that blank cassette sales far outstripped both blank open-reel and blank 8-track sales from about 1970 onward. At that time, however, sales of recorded cassettes were quite low in comparison, a fact that makes the cassette's rise in the early 1980s seem more dramatic. In 1974, recorded cassette sales in the United States were still only about 16 percent of 8-track sales. By comparison, sales of recorded open-reel tapes that year were about 0.005 percent of 8-track sales, or 0.02 percent of cassette sales. Eight-tracks, which had held over 82 percent of the tape market in 1976, plunged to about 47 percent by 1979, the year before record companies announced their intentions to drop the format. From that time on, the path was clear for the cassette's ascendancy (Morton, *Off the Record*, 2000, 166).

While the cassette hardly needed a boost at this point, it got one anyway. In 1979, the Japanese electronics company Sony, which had scored a hit several years earlier with its Betamax videocassette recorder, introduced a small, battery-operated cassette player in Japan. The device, called the Walkman, was not fundamentally different than the pocket-size cassette players it had offered for several years, and except for the provision for stereo playback, it was not strikingly different than the handheld cassette dictation machines it had been selling for some time. The recorders were sold the following year in New York and other key American markets. Priced at about $150, the little tape players were far from inexpensive, yet they struck a chord with fashion-conscious New Yorkers, and soon they were the focus of stories in nationally read newspapers such as the *New York Times*.

As sales of the original Walkman began exceed expectations, Sony ratcheted up its publicity effort to support a new model in 1981, the WM-2, which was the first Walkman that most people outside New York or Tokyo ever saw. That year the Aiwa Corporation offered a similar device that could also make recordings, and Sony followed suit a few years later, but it was players, not recorders, which would take the largest share of the market. That fact more than any other proved what marketers already knew—that portable electronics was primarily about music listening, not the making of recordings. The great economic impact of the Walkman and its imitators in the 1980s was matched by its influence on listening habits. Music, which had earlier moved from the living room to the automobile, was now carried on the human body wherever it went.

At the time, critics predicted that the rise of "personal stereos" meant that people, particularly urbanites, would become more isolated and less social. The aural environments that people created for themselves when they used their personal stereos with headphones not only drowned out external sounds but also discouraged social interaction and conversation. A person on the street with a Walkman and a pair of dark sunglasses seemed virtually unapproachable.

The bright side of the Walkman revolution, as critics saw it, was that it helped diminish the success of another major tape-related product of the early 1980s: the boombox. Like portable radios, battery-operated tape players with built-in loudspeakers had been around since at least the early 1960s. They had been growing in popularity since the introduction of portable Muntz Stereo-Pak and later 8-track cartridge players, and had then taken off when cassette recorder sales to teens and children began rising in the later 1960s. During the 1970s, more expensive, feature-laden, battery-operated portables had been appearing, and by the middle of the decade the top-of-the-line cassette portables featured multiple loudspeakers, powerful amplifiers, and stereophonic sound. They had Dolby noise reduction, the ability to play the latest chromium-dioxide tape formulations, and incorporated an AM-FM radio, a tape player, and the ability to accept the input of a phonograph. While still portable, these large machines were complete stereo systems in one compact package.

Panasonic scored a big hit in 1978 with a heavily promoted line of full-featured and pricey (often $150 or more) units. A large market began to emerge around the world for these devices, but they were notoriously associated with American urban youth. These Americans seemed attracted to the boomboxes not only for their high-quality sound and portability but also for the sheer volume of noise that they could generate. African Americans, Hispanics, and other groups in large cities latched onto the idea of playing music loudly in public places as much for their own entertainment as to attract attention and make a political statement through the music they blared.

A backlash followed. Because the technology became so closely associated with urban African Americans and their music (rap music had become popular in the early 1980s), hateful new names emerged, such as "ghetto blaster," "nigger box," or even the absurd-sounding "third world briefcase." Manufacturers tacitly acknowledged these slurs when they began to advertise the devices as "boomboxes" and "blasters" later in the decade, but the original connotations faded. Cities were quick to outlaw the use of boomboxes (unless headphones were also used) in public places such as subways, a move that some interpreted as noise abatement but others as suppression of free speech.

Despite this controversy, the boombox was an enormous commercial success. By the late 1980s the boombox-style system had not only become one of the most popular forms of mobile audio gear, but it had also taken its place in the home, replacing the earlier all-in-one radio-phonograph combination as the best-selling type of living room system (Morton, "History," 1999, 50).

THE COMING OF THE CD

The mid-1980s saw sales of recorded cassettes exceeding those of the LP record. By 1986, almost three times as many recorded cassettes were sold in the United States as discs. As in the case of the 8-track, record companies did not wait for LP sales to decline to nothing, but eliminated them from their catalogs preemptively between 1988 and 1990. After more than a century in production in various forms, the phonograph was finally obsolete (or so it appeared for the moment). There had been few who regretted the demise of the 8-track, and fewer still who pined for the days of the 78 rpm disc, but the elimination of the LP struck many as being premature. Album sales had been strong through most of the 1970s, and many people had amassed considerable record collections. Despite the cassette's ongoing technical improvement, many people preferred to purchase LPs for use in the home, and used cassettes on the road or in portable players. However, it was also true that the masses of consumers bought recorded cassettes primarily and played them on boomboxes or inexpensive, all-in-one home systems that increasingly were not supplied with a phonograph. Yet the cassette continued to be plagued by quality problems. Record companies were apparently unable to find ways to improve the sound quality of mass-produced cassette recordings to the point where they matched the LP in all respects, and some artists objected to releasing albums only on formats they considered inferior.

Nonetheless, when the record companies stopped releasing albums, there was little that consumers could do other than purchase a compact disc player. The compact disc would become the Phillips Corporation's first great commercial success since the introduction of the Compact Cassette in the early 1960s. Like the 8-track before it, the compact disc had been preceded by a technology intended for video rather than sound. Phillips, along with several other electronics firms in the 1970s, was experimenting with ways to record analog television signals optically and reproduce them by scanning the medium with a beam of light.

Phillips teamed up with the American company MCA, the record manufacturing arm of the Universal Pictures movie studios. MCA had in turn purchased the patents of a small engineering firm in California, whose founders had developed some basic ideas for an optical videodisc system for television or motion picture images. MCA's first prototype disc player was demonstrated in 1972. Because the invention was so similar to what Phillips had arrived at independently, the two firms agreed to cooperate. The combined system, as demonstrated again in 1977, used a 12-inch plastic disc, spinning at about 1,800 rpm and holding 25 minutes of audio and video. The recording was made by using the signal from a video amplifier to modulate a laser. The laser (then an expensive and delicate device) was focused on an aluminum master disc and burned a spiral path of minute pits into the disc, representing the undulations of the original signal. Discs were duplicated by stamping, similar to the process used for LP records, and pressed into clear vinyl. The duplicates, labeled and sold to the public, were read by a second, less powerful laser, which shone on the pits. A wavering light reflected from the surface of the disc was detected by an electronic optical sensor and converted into an electrical signal, which could be amplified, processed, and fed to a television receiver. The production system was first sold in Atlanta, Georgia, in 1978 (Morton, "Disc Television Recording," forthcoming).

In the midst of this, the Sony Corporation introduced its Betamax videocassette recorder in 1976, followed by the introduction of the competing VHS videocassette recorder the following year. While the Betamax would eventually lose this battle, VHS became the standard household video system, revolutionizing the way people watched television. The impact of this upstart technology on the videodisc was decidedly negative. While MCA's system faded quickly, other companies such as Pioneer introduced improved laser videodisc technologies several times over the course of the 1980s. Each of them failed or gained only a small following.

THE ARRIVAL OF DIGITAL AUDIO

Phillips, however, was working on a new plan. The same videodisc technology used for analog television signals would work for digital audio. Just as television signals consist of brief electromagnetic pulses, converted to pits of varying sizes for recording video, digitized audio also consists of pulses that could be translated into pits of regular size on the surface of the disc.

Digitized audio had already been in existence for years before Phillips etched it onto laser discs. English inventor Alec Reeves invented pulse code modulation (PCM) in the 1930s, a technique for converting an ordinary audio signal to a digital signal. Later, telephone companies such as AT&T developed PCM equipment that allowed them to pack multiple voice conversations onto a single wire, leading to greater efficiency in the use of long-distance telephone cables. Yet the recording of digital audio was rarely undertaken before the 1960s.

The NHK Technical Research Laboratory in Japan had demonstrated an early studio-type digital audio recorder in 1967, sampling an audio signal using PCM techniques and recording the resulting data as spots of magnetic flux on a commercial videotape. The technical requirements of PCM audio recording are so similar to the requirements for television recording that the two technologies have subsequently evolved together. Two years after the NHK demonstration, Sony Corporation introduced a commercial digital audio recorder for studio mastering purposes. With commercially available studio recorders now available, several firms looked into providing a way to bring digital recording to consumers. The first digital audio recorder to reach the market was built by the Sony Corporation. It consisted of a Betamax VCR coupled to an external "black box" device to convert an incoming signal to digital form, and to convert the PCM-encoded signal on the tape to analog form that could be handled by an external audio amplifier.

However, Sony's involvement in laser videodiscs at about this same time led engineers to consider an audio-only version. Using the same type of PCM encoding, but recording it onto the type of medium used for making laserdiscs, Sony demonstrated a prototype audio system in Europe and Japan in 1979. These demonstrations led to a cooperative agreement with Phillips, whose engineers were also working on such a system themselves. The successful development of the compact disc, as it was to be called, relied only in part on established digital recording and videodisc technologies. It also required the invention of ways to mass-produce certain key components at low cost for use in the consumer players. The complex electronics of the original Sony PCM coder/decoder of 1977, for example, had to be recreated as a set of small, inexpensive integrated circuit chips. Further, a laser device cheap enough to install in a consumer disc player and capable of being mass-produced was not available until about 1981. With these key technical hurdles out of the way, the two companies unveiled the CD in 1982.

SELLING THE CD

The compact disc was hardly an instant success. The price of the first Sony players sold in the United States was about $2,000, and the discs themselves were priced at $12–$15. But after just a year, it was possible to purchase a second-generation player for under $700, falling to less than half that after another year. Consumers nonetheless resented the high prices for the discs, and the claims of the record companies that mass production would greatly

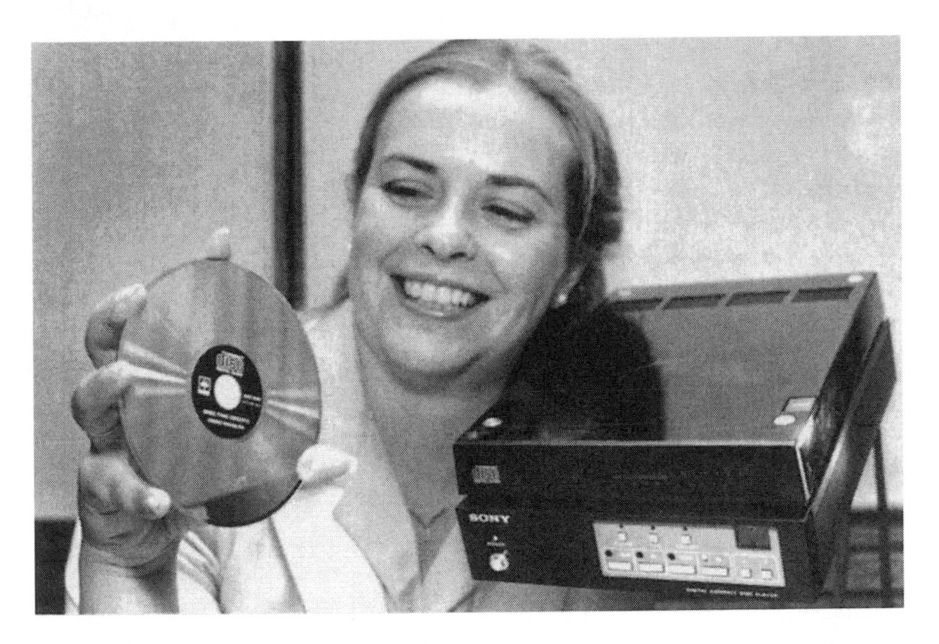

The Sony compact disc makes its debut at the Chicago Consumer Electronics Show, 1982. CORBIS.

lower the cost were not made good. However, the appeal of the CD was strong. Its ease of use, small size, and durability combined with excellent audio quality made it a natural successor to the cassette tape for use in both home and mobile systems. Because the CD player could not record audio, sales of blank cassettes and recorders remained strong for many years, as consumers continued the practice of copying from purchased (or borrowed) albums on CD. It was not until 1994 that CDs first outsold cassettes worldwide. Yet outside the United States and Western Europe, the cassette remained strong much longer. Recorded cassette sales by 2000 accounted for about 15 percent of North American sales, but over 35 percent of sales worldwide (IFPI 1999).

THE COMING CHANGES

While the compact disc did not suit everyone perfectly, it did seem to be the last word in high fidelity. The century-long quest to improve sound had seemingly ended. Audio engineers had established measurement-based standards to define high fidelity as early as the late 1940s, and these same standards are used to judge new audio formats today. The ideal recording medium under those standards would capture the entire range of audible frequencies from about 20 to 20,000 Hz, and would do so with no measurable noise or distortion. The compact disc approached that ideal so closely that improvement hardly seemed necessary, except perhaps to a stalwart few who insisted that the discs still were not quite good enough. Given that, for most people, there was nothing more to ask of recording technology, what could possibly come next? But the last decade of the twentieth century did see another revolution in recording technology. The coming change was not a story of radical technical innovation, but instead a radically shifting set of expectations. Just as the rise of car tape players, the Walkman, and the boombox had reflected a revolution in the way music was heard, the coming changes in recording technology would reflect changes in the way consumers obtained and enjoyed music.

16

Record Companies versus the World

ON EDISON'S TURF

It is common today to hear news about the recording industry's battles against the unauthorized copying of music and videos. Since the late 1990s, those struggles have been centered on the Internet and digital music (topics discussed in the next chapter). Yet the problem of unauthorized duplication has been around since the nineteenth century, changing in its nature and form alongside changes in the technologies of recording.

PHONOGRAPH COPYING

As historians have noted, while Thomas Edison invented the phonograph in 1878, he failed to commercialize it promptly. This gave his competitors a chance to extract his best ideas and start improving them. Thus, the second generation of recording technologies was in some sense an unauthorized version of the original. Those invading Edison's territory included the resourceful inventors at the Volta Laboratory in Washington, D.C., who introduced the wax cylinder graphophone in the 1880s. Edison returned to the phonograph and attempted to leapfrog their innovations by introducing his own improved phonograph, which incorporated many of the graphophone's best ideas and added a few more.

Copying the phonograph was made much easier through the publication of how-to books on the subject. In 1892, an author named W. Gillett published *The Phonograph and How to Construct It*, a book that led to widespread copying of Edison's invention and encouraged several inventors to patent their own versions of it. In fact, throughout the history of sound recording there have been numerous claims and counterclaims about the illegal use of patented designs or other intellectual property related to recorder technology.

But it is primarily the unauthorized duplication of the content of phonograph *records* that concerns us here. Edison and others complained about the practice (apparently of some licensees) of taking the music cylinders that his company made in New Jersey, making multiple copies of them, and selling them under a different brand name. This practice denied Edison's company of its share of the profits from the sale of cylinders. In later years, unauthorized music copying came to be known as either "piracy" if the unauthorized recordings were sold under a new brand name, or "counterfeiting" if the unauthorized recordings were manufactured so that they carried the original brand or looked like the original product.

The copying process for records was never simple, but in the early days of the phonograph it involved a minimal investment in equipment. Commercial records in the 1890s were made by mechanically transferring the recording to a blank cylinder using a delicate "pantograph" mechanism, which traced the shape of the original groove mechanically and inscribed a matching groove on the blank. The pantograph method disappeared around 1900 as Edison and others introduced cylinders made by a molding process. Copies made by molding were so much better sounding than the pantograph copies that the latter could not compete. Molding was also more capital intensive than the earlier process, and so it presented a high economic barrier to would-be pirates and counterfeiters.

Discs, introduced in the late 1890s, could also be copied by the pantograph method, but by the time they were introduced the pantograph was obsolete. Because the equipment necessary to mold discs was elaborate and expensive, it helped exclude the small-time pirates. Subsequently, most illegal copying was done by record manufacturing companies, who sometimes produced their copies in addition to making legitimate products.

Disc pirates still used a store-bought commercial record as the basis of their copies, a fact that always resulted in somewhat lower sound quality for the pirated versions. The details of the process are not well documented, and it is not clear for example whether a new wax original was created, or whether the commercial disc itself was plated to make the stampers. The rest of the process would have been the same as with a legitimate record. Independent record duplicators, firms that manufactured recordings for large and small record companies under contract, or that produced small batches of transcription records, were well prepared to undertake such illegal ventures. Their customers included record retailers and distributors looking to make an easy buck. There was also considerable copying of American and European records done elsewhere in the world, where Western copyright laws could not be effectively enforced.

BROADCAST RECORDINGS

When radio came along in the 1920s, it introduced a new source of stealable material. Broadcasts, under ideal conditions, had relatively high-quality sound and often featured well-known artists. The most valuable broadcast music emanated from the major networks, NBC, CBS, and Mutual, who regularly broadcast feature musical programs. Around the same time, disc recorders were being improved, and "electrical" recorders came into wide use in radio stations by the late 1930s. These recorders, though expensive and difficult to use, were capable of making high-quality recordings off the air or, even better, directly from the network feed cables that the telephone company provided to radio stations. However, it was unusual for stations to make unauthorized recordings of network programs until the advent of the acetate plastic, "instantaneous" transcription disc, which was lower in cost. Many program recordings that were later pirated were originally made for legitimate archival purposes on acetate discs. Later—sometimes decades later—they were released by independent record companies, even though the recordings were not licensed for duplication. In most cases, the sources of the recordings were not identified, making them difficult to trace. By capturing radio performances of big-name performers or bands, record pirates got access to high-quality content not available on authorized records. The resulting discs were sold by mail order or sometimes by ordinary record retailers, often at a lower price than comparable legitimate records.

THE JERSEY CONNECTION

A major federal hearing on the issue of record piracy and counterfeiting took place in 1962. Recording industry lawyers presented evidence showing how organized crime in New Jersey had operated a pressing plant in the

1950s for the purpose of counterfeiting records, which were then sold to jukebox operators or record retailers. New York–based record dealer Sam Goody was charged with selling such records, and was convicted of having purchased counterfeit copies of a Glenn Miller recording captured from a radio broadcast, selling the discs at a discount in his New York stores.

With the advent of the LP after 1948, there were a great many "greatest hits" and other compilation-type albums, consisting of the older hits of popular artists. Along with the new disc came a new wave of piracy, as the incidence of unauthorized compilations taken from old network programs or dubbed from individual 78 rpm discs increased. The new tape-recording technology contributed to this trend, since tape could easily be used to copy commercial recordings, edit them into an appealing compilation, and transfer them to a new master record. One unexpected result of the transition to the LP was that for a brief period in the late 1940s it became less expensive for potential record pirates to get into the business. The advent of the LP prompted legitimate record manufacturers to replace their older record stampers. By purchasing these used record stampers at low cost, counterfeiters were able to enter the market more easily.

Partly in response to this rash of LP piracy, the Recording Industry Association of America (RIAA) was formed in 1951. Since that time, the RIAA has become involved in numerous other activities, such as the certification of "gold" records and the establishment of technical standards for sound recording, but the group's antipiracy lobbying efforts on behalf of the record companies continue to this day. According to the RIAA, the problem of unauthorized copying was severe even half a century ago. The 1962 congressional hearings, for example, revealed that the record industry was by one estimate denied $20 million in sales during 1960 alone due to record counterfeiting. Laws passed in 1962 established a fine of up to $10,000 and a year in prison for record retailers who dealt in counterfeits. Another hearing recommended in 1965 the toughening of federal penalties for record pirates.

BOOTLEGGING

With the advent of tape recorders after 1945, record companies gave a mixed message to consumers about the new technology's potential for copying commercial recordings. A few industry individuals publicly expressed concerns about the use of home tape recorders to make unauthorized copies. It was not made clear, however, how record companies could be hurt by such activity, as long as consumers did not begin selling the tapes they made.

By the 1960s, high-quality home tape recorders could make recordings nearly as good as ones made by professional machines. Consumer electronics manufacturers were also introducing "semiprofessional" open-reel recorders to serve a small but growing market consisting of musicians and others who wanted to make their own recordings. A combination of improved quality and increased familiarity with recording techniques led to all sorts of experimentation, and soon resulted in a new type of unauthorized recording: the bootleg.

Unlike a pirate or a counterfeiter, the bootlegger captures and sometimes sells copies of a recording or performance that has never been officially released or broadcast in any form. Bootleg recordings generally fell into two categories. The first consisted of studio outtakes, including performances and often the between-song chatter, plus other unreleased bits and pieces that accumulated in the recording sessions of the late 1960s. These unauthorized studio-based recordings began to find their way to consumers on quarter-inch tape, LP records, and eventually cassettes. Although awareness of such bootlegs is generally agreed to have emerged in 1969 with the marketing of an unauthorized album of Bob Dylan tunes called *Great White Wonder*, the practice probably originated somewhat earlier. The second form of bootleg was also related to the use of the tape recorder, but in a different way. Beginning in the 1960s, numerous rock concerts and other performances were captured on tape and bootlegged, either by fans holding battery-operated portables near the loudspeakers, or by sound system technicians who simply patched into the mixing board.

EIGHT-TRACKS AND PIRATING

The rise of the Stereo 8 tape format in the late 1960s made both piracy and home copying a much more important phenomenon. Small-time counterfeiters were able to copy albums inexpensively in small batches onto 8-track tape using simple equipment or even home recording decks. For larger operations, commercial 8-track duplicating equipment was also relatively inexpensive compared to disc stamping machinery, and the per-unit cost of copies was low. For counterfeiters who bothered to try to duplicate closely the legitimate product, making fake 8-track tapes required only the tape itself, its generic plastic housing, and one printed label, compared to the LP, which demanded not only the disc but two labels, a liner sleeve (sometimes printed with lyrics), and a printed cardboard outer sleeve in order to appear to be the legitimate product.

In some sense, record companies inadvertently contributed to the rise of 8-track piracy by treating the new format as a sort of "second-class citizen." Some record labels farmed out 8-track duplication to third-party firms rather than bothering with it themselves. Few of these firms were tempted to become counterfeiters, but the tape duplication industry stimulated demand for low-cost duplicating equipment, which was then used by others to make counterfeits. Legitimate 8-track duplication was dominated by a few large firms, but there were hundreds of small firms in the United States and around the world, some of which clearly became counterfeiters. Easily available tape duplicators, the lack of company-owned duplicating plants, and a product that was simple to copy made it easier for counterfeiters to get into the business and avoid detection.

CASSETTES

The market for recorded 8-track tapes waned as cassette (the term "cassette" became generic as early as 1975) technology improved in the late 1970s, but the cassette created enormous problems for the record companies. They did come to appreciate the economics of selling music on cassette, discovering that cassettes cost less to produce than an LP or an 8-track while selling for as much or more. Yet the cassette was first and foremost a technology for recording, not just listening. While the 8-track's popularity had been founded on sales of recorded tapes, the cassette had risen to prominence first as a home recording medium, and only later did it gain acceptance as a way to enjoy recorded albums. Although record companies by the mid-1980s shifted from indifference to enthusiasm for the the cassette as a replacement for the aging LP, consumers had already by this time embraced the practice of using tape technology to make recordings at home. The cassette was a two-edged sword that delivered short-term profits and threatened long-term disaster.

Nearly all cassette equipment sold in the United States included recording circuits (which was not usually the case with the earlier 8-track system), so that virtually any machine could be used to make a copy from a phonograph, the radio, or another cassette recorder. As they became familiar with cassette-recording techniques, consumers found it fairly simple to duplicate either LPs or recorded cassettes using inexpensive home equipment.

Cassettes were just as appealing to the professional counterfeiters. They were as simple and inexpensive to counterfeit as the 8-track had been, yet the cassette market was larger. An important difference between the two formats was the cassette's greater portability and its diminutive size, which made it easier to design truly efficient, battery-operated portable recorder/players. After the Walkman and boombox had captured the fancy of Americans and Europeans in the early 1980s, the introduction of a growing variety of cheap, portable recorders helped give the cassette a global appeal that the 8-track had never attained. Soon the cassette was the medium of choice for everyone, legitimate consumers and counterfeiters alike, all around the world, giving piracy and counterfeiting an unprecedented economic impact on the recording industry.

ANTIPIRACY LEGISLATION

The anticounterfeiting laws of the early 1960s had apparently been ineffective, and new laws were passed in 1972 to strengthen the powers of the record companies in both civil and criminal courts. Part of the problem with the earlier legislation was that, in the United States at least, there were no copyright laws that protected the content of recordings themselves. The only laws that could be enforced in piracy and counterfeiting cases were related to the rights of songwriters and music publishers. In *Duchess Music Corp. vs. Stern* (1972), for example, a defendant was convicted of making 25,000 copies of a compilation tape and offering them for sale. She was sued by a record company not for the act of copying but for failure to pay the compulsory license fee of 2 cents per song.

Only in 1971 and 1972 was federal copyright protection extended to musical performances captured on recordings, giving record companies the protection already enjoyed by music publishers. In 1971, most Western European nations enacted international agreements to try to suppress the trade in counterfeits, an agreement called the Universal Copyright Convention, and this was ratified in the United States in 1974. The U.S. Supreme Court in 1973 upheld the rights of states to prosecute music pirates and counterfeiters, although at that time few states had their own antiduplication laws, so unless the trade in unauthorized recordings was conducted across state lines (thereby violating federal laws), it was still difficult to prosecute counterfeiters. Still, the new laws encouraged the record companies to step up efforts against large-scale offenders both in the United States and abroad. As part of this heightened effort, in 1981 Sam Goody was again in the news, after FBI agents raided his namesake company, Sam Goody Inc., and seized evidence that the company had sold millions of dollars in bogus records.

CONSUMERS AND UNAUTHORIZED RERECORDING

While it was becoming easier to convict major offenders, a big question that remained was what to do about home copying. Because few consumers tried to sell the recordings they made, charges that they had violated copyright laws seemed untenable. Yet by this time, many middle-class homes were equipped with multiple audio systems, and it became common for consumers to purchase records and make copies for use on other systems, such as in-car players or boomboxes. In 1978, the RIAA claimed that its members lost $500 million each year because of home taping on analog cassettes. Record companies began to pressure industry associations and governments to impose taxes on blank cassettes and recorders in order to make up for their perceived losses due to the unauthorized copying of commercial releases.

These efforts grew fervent in the early 1980s as consumer tape-duplicating equipment appeared in the form of "dual-well" recorders. These recorders could be used to quickly and cheaply copy recorded cassettes. Yet the courts proved stubbornly resistant to the idea that home taping hurt the record industry. A major setback from the industry's perspective was to come in 1984, following a suit (first filed in 1976) known as the "Betamax" case (*Song Corp. v. Universal City Studios*, 1984). In it, an individual had been identified and sued by Universal Studios as a test case. His crime: recording a television program off the air on a Betamax VCR. After years of litigation, the U.S. Supreme Court finally ruled that home videotaping in itself did not represent a copyright violation. The ruling was interpreted to extend to home audiotaping as well.

Record companies subsequently adopted a new strategy, attempting to pressure U.S. lawmakers into legislating changes in the design of audio recorders. Their efforts resulted in the voluntary inclusion in every new VCR of an anticopying circuit developed by the Macrovision Corporation (incidentally, the Macrovision system did not affect the audio portion of a videocassette recording). But no one seemed able to devise a practical way to prevent consumers from duplicating copyrighted audio recordings short of removing the recording features of tape equipment altogether.

DIGITAL AUDIO AND THE PIRATE'S PARADISE

The story of consumer digital audio is still unfolding. The very first home digital audio–recording equipment appeared in the late 1970s, but the technology was not well known until the advent of the playback-only digital compact disc in 1982. The technical press speculated that the CD, touted as an improvement in sound quality, was being pushed by the record companies not because of its aural qualities but because it was difficult to copy, and this was mentioned publicly by industry executives as one of the medium's desirable features. In fact, the CD's clean sound could be used to make high-quality analog recordings onto cassette with ease, but the players were incapable of making either analog or digital copies onto another compact disc, simply because manufacturers were not yet offering a home CD recorder.

Home digital recorders of another type became available in 1987 with the introduction of digital audio tape, the brand name for an early digital cassette. The DAT, as it was popularly known, became one of the most controversial new technologies of the late twentieth century. Even before it was introduced, record companies argued that the machine, which was projected to cost around $1,000, lowered the economic barriers for mass copying and would breed a new generation of counterfeiters. They pushed hard for restrictions on the ability of the machines to make digital copies from an external digital source, such as a second DAT recorder or a CD player. Interestingly, some of the first consumer DAT recorders were offered by Sony, which had maintained a business relationship with CBS records since 1968. This seeming contradiction was reflected in the public statements of many electronics companies, who wanted to cooperate with the recording industry but did not want to fall behind in offering the next generation of products.

But by this time, representatives of the record industry were prepared to do battle. They lobbied hard to pass the U.S. Audio Home Recording Act of 1992, which mandated that all digital recorders sold in the United States would contain electronic circuits to prevent or degrade "serial" copies (or copies of copies) of digital recordings just as VCRs had provisions to prevent copying from VCR to VCR. Within a few years, international agreements were signed that also prohibited the sale of devices intended to defeat the electronic "locks" intended to prevent copying. But because making analog copies was not restricted, the bill provided small royalties to songwriters, recording artists, and record companies to make up sales lost to consumer copying.

Meanwhile, various companies were developing a number of different systems to try to prevent digital-to-digital copying, and this was seen as the key to preventing mass copying by pirates (digital-to-analog copying is not affected). Early Sony DAT recorders of the late 1980s used something called the Social Gathering System. Later digital audiotape and digital disc

Publicly destroying counterfeit CDs, movies, and cassettes seized in Bangkok, 2001. CORBIS.

recorders used the Serial Copy Management System (SCMS). Both of these systems added data along with the digitized music that permitted or restricted copying when the recorded data was read by a player. In the SCMS system, the state of a single bit of information determined whether a recorder receiving a digital data-input stream would work properly or remain locked in the "pause" mode. Yet none of these systems initially seemed very important, because the public had failed to purchase digital tape recorders in significant quantities. It appeared that for the time being, consumers would stick with the playback-only CD and continue to use cassettes to make their recordings.

What at first seemed like a success in the industry's ability to control piracy lasted only four years. By 1996, the first CD-R ("R" for recordable) technology appeared. CD-R devices were not subject to the Recording Act restrictions because they were usually part of a personal computer rather than a device intended solely for recording music. In fact, the CD-R was not limited to audio, since it could record any kind of computer data with equal facility. As the cost of recordable CD computer drives and media came down, their popularity rose. While Microsoft and some other software companies voluntarily wrote copy-protection restrictions into their CD-recording software, the Internet helped spread other programs that allowed users to "rip" digital copies of copyrighted works and "burn" them onto audio CDs with ease.

From the recording industry's perspective, even greater problems arose when consumers began to make conversions to the newer MPEG-3, or MP3, format. MP3 files, often converted from the audio files on ordinary CDs, began showing up on Internet newsgroups, where they could be accessed by any of the millions of Internet users worldwide. Unlike the earlier digital format used to make audio CDs, even "legal" CD-burning software does not usually place any restrictions on serial copies of MP3 files. Once an individual has converted the content of a commercial CD into MP3 format and distributed it on the Internet, it is a simple matter for others to copy it.

Home copying of copyrighted recordings had taken on an entirely new character. In an early effort to strike back, the RIAA brought suit against the makers of the Diamond Rio portable MP3 player in 1993, but were unsuccessful in forcing its makers to pay penalties or prevent its users from copying music files from their computers to the device. More trouble was on the way. By 1999, a World Wide Web–based file-sharing service called Napster revolutionized the distribution of unauthorized music files by giving even the least savvy computer users a simple way to exchange their MP3-encoded collections. Under attack, Napster drained its resources on legal defense and then sold its assets to the publishing giant Bertelsmann AG, which then subsequently let Napster slip into bankruptcy in 2002 (it returned to life in a different form the next year). In its wake, imitators sprang up and remain in operation today, and users began employing non-Web means to distribute MP3s, such as Internet relay chat, or IRC.

POST-NAPSTER

With the rapidly falling cost of CD recorders after about 2000, many consumers began using post-Napster file-sharing services to compile new collections of pirated recordings. However, upcoming generations of computers will most likely reverse this trend. Newer anticopying schemes treat all digital music files—copyrighted or not—as if they were copyright protected. If these protections succeed, it will probably not be possible to compile audio or video recordings from one's own collection, and this is a sharp break from earlier policies. Most likely, consumers will either illegally circumvent these restrictions or will give up recording altogether as content producers become able to deliver more and more products on demand to a variety of devices.

17

Online Music and the Future of Listening

CONTINUAL REVOLUTION

Since its invention, sound recording technology has continually changed and evolved. The drive to improve sound quality (even though it is difficult for everyone to agree what good quality is) has been a constant factor in this evolution, usually leading to small innovations that were slowly incorporated into studio practices, record manufacturing techniques, or consumer technologies. There have also been inventions that were more revolutionary in nature. The phonograph itself was a revolutionary invention, transforming the scientific field of sound recording (using the phonoautograph) into the commercial field of sound recording and sound reproducing. Magnetic and optical recorders were other revolutionary changes, their success requiring significant transformations in the ways recordings were made and used. This book has argued that certain innovations were revolutionary not only because a new technology was involved, but also because the new technology inspired (or was inspired by) new ways of recording or listening. The commercial success of the disc gramophone is an example of this. While Edison and others had already established the sound recording on cylinder as a form of entertainment, the disc format was better suited to capitalize on the fad for purchasing and collecting recorded music, and the disc's corporate sponsors did a better job of giving the public the music it wanted. Similarly, tape recorders were a relatively insignificant consumer product until new corporate sponsors arrived in the 1960s, repackaging tape as a way to make music portable, and marketing it to a new segment of the population.

THE CD AS INNOVATION

Viewing the history of recording in this way makes it necessary to argue that the compact disc and other forms of digital recording were not themselves revolutionary in nature. That statement runs counter to most of the marketing "hype" that has accompanied the CD since its introduction in the early 1980s. CD player manufacturers, record companies, and music journalists have generated a myth that digital recording technology is so new, so "cutting-edge," that it must be revolutionary. It is true that at the time of its introduction and even twenty years later, it represented the highest of hi-fi. Unlike the LP, it did not require operating-room cleanliness to handle and store without degrading its sound. Unlike the cassette tape that preceded it, the CD never stretched, broke, or got "eaten" by its player.

However, from another point of view, the CD can be seen as offering merely incremental improvements over the previous generations of technology. It was from the beginning a highly portable medium, continuing the trend started by the car tape player, the Walkman, and the boombox. Walkman-style CD portables, CD boomboxes, and car CD players appeared within a year of the home CD player's introduction. The CD's small size and durability made it a good—although not perfect—medium for portable listening. In other respects, the CD was much like the LP, holding about the same amount of music and selling for a price that was comparable, if usually a bit higher. In technical terms, the CD was arguably an improvement over the LP. Compared with the LP, the CD could record a wider band of frequencies. The CD's bandwidth encompassed the very lowest 20 Hz or so at the bottom end of the hearing range, and the several thousand Hz at the top end of the audible range. While this could be seen as a major improvement, it is worth noting that most people simply cannot hear the frequencies above about 15,000 Hz. Ideally, human hearing extends up to about 20,000 Hz, but the occurrence of humans who can hear the entire range, especially among those over thirty years old, is at least as rare as the number of people with perfect eyesight. The CD's strongest improvement over the LP was in the area of what audio engineers call dynamic range, which is the difference between the loudest sound and the softest sound that it is possible to record. The lower limit, which is probably the most important of the two, equals the background noise level; on an LP, the background noise is the unavoidable byproduct of the stylus dragging along the surface of the vinyl, added to the hiss of the master tape, which is transferred to the disc during mastering. The CD, especially when coupled with the new generation of digital studio recorders, offered a dynamic range that was considerably better than that of an LP.

But did those improvements in fidelity constitute a revolution? People tended to use the CD and its partner, the CD-R, in the same ways that they used the LP and the cassette for the recording and playback of music. From an anthropological or sociological point of view, there was little to distinguish the new digital technology in terms of its role in society. In the last years of the twentieth century, however, the CD and digital recording technology converged and became part of a development that was truly revolutionary: online music.

TRANSMITTING DIGITALLY

The controversial technologies for storing and sharing music over computer networks began as a search for ways to "compress" digital audio and video data. Pulse code modulation or PCM was from the early 1980s the standard way of converting analog audio and video signals into digital form and storing them on tape, CD, or other media. However, the designers of the original PCM audio recorders, such as Sony and Phillips, were very much concerned with preserving high fidelity. The system used to encode CD audio, for example, captures and records data at the rate of over 175,000 bytes per second of audio, or about 10 megabytes to record a minute of stereo music. That high rate works well for an audio CD, where disc space is not at a premium and masses of data can be transferred within integrated circuits, but it presented problems when engineers began looking for ways to send high quality digital audio over telephone lines or over the airwaves. Engineers in the early 1980s developed a way to compress video data for transmission over telephone lines, intending to use this technology with the next generation of videophones. This began a line of research that resulted in new ways to transmit and store both audio and video.

In 1987, the German nonprofit research consortium Fraunhofer Gesselschaft, through its laboratories at the University of Erlangen, began working on a project to compress the data stream of a proposed digital radio broadcast. The compression of digital audio data had been practiced since at least the early 1970s in the telephone industry, where telephone companies digitized an analog signal, then used computer algorithms to mathematically reduce the volume of data. For example, if part of the data stream happens to consist of a long run of the same digit, say a string of binary zeros, it is possible to substitute a shorter piece of data that contains just a description of the data and the length of the run. A coder at the transmitting end is matched with a decoder at the receiving end that contains circuits or software instructions to enable this kind of compressed communication. The result is that it is possible to accomplish things such as transmitting wide-band digital sound (such as full-range, high-fidelity audio) through a medium that is designed to transmit narrow-bandwidth analog audio (such as a voice telephone line).

The digital radio project at Erlangen sought to compress audio data so that the broadcast would require less bandwidth, which was important because of the scarcity of available frequencies and the limitations of existing bandwidth allocations. Researchers led by the Erlangen laboratory's leader, engineering professor Dieter Seitzer, developed an improved mathematical method that could reduce the bandwidth of a digital audio data stream by a factor of twelve with only a minimal loss of sound quality. It did so by analyzing the audio data and using sophisticated algorithms to remove redundant or irrelevant parts of the signal. For example, if a loud sound like a crashing drum drowns out quieter sounds in a musical passage, then those quieter sounds are irrelevant from the listener's standpoint. Removing them reduces the overall amount of digital data without noticeably distorting the signal. Seitzer's team proposed several incremental levels or "layers" of data reduction, corresponding to increasingly greater levels of distortion of the original signal. The best data compression that they could achieve before the typical listener began to notice the difference was between 1:10 and 1:12; this became "layer III" compression.

Fraunhofer Gesselschaft received a patent for the compression algorithm in 1989, but by this time there was growing interest in technologies such as digital telephones, cellular digital telephones, videophones, and videodiscs. Knowing that it could be applied to audio and video, and hoping that manufacturers would adopt the new standard for any or all of these new technologies, they submitted it to a committee called the Moving Pictures Experts Group (MPEG), created in 1988. This committee, originally consisting of twenty-five experts in the field, was jointly sponsored by two international standards-setting bodies, the International Standards Organization (ISO) and the International Electrotechnical Commission (IEC). The original aims of the MPEG were to evaluate various proposals for compressing high-quality digital video and audio for transmission or storage on a compact disc, and to standardize one or more of the standards in order to push forward the devel-

The future of listening? Apple Computer's founder Steve Jobs holds the Apple MP3 player, 2001. CORBIS.

opment of consumer videodisc technologies. By 1992, the group published its first set of compression standards, which it called MPEG-1. For the audio that accompanied video, the group recommended the patented Fraunhofer algorithm, which became "layer 3" of MPEG-1. This cumbersome term, MPEG-1 layer 3, was eventually shortened to MPEG-3, then simply MP3. That year, Sony Corporation introduced its Minidisc, a smaller, recordable audio CD using MP3 encoding, but the format failed to become popular outside certain niche markets.

THE INTERNET AND MUSIC

In the meantime, the use of the personal computer and the Internet was exploding. The distribution of digital music on the Internet started about a year after the publication of the MPEG standard. In November 1993, one of the earliest online music sources appeared in the form of the Internet Underground Music Archive, created by Jeff Patterson, Robert Lord, and others at the University of California, Santa Cruz. A bit later, in 1994, a seminal Internet discussion group began, called the MPEG-3 Audio Consortium, which itself used the Internet to link people who were interested in the new medium. As the number of enthusiasts grew, privately sponsored archives of songs began to appear as file transfer protocol (FTP) sites, accessible through pre–World Wide Web search engines and file-transfer software packages. Such FTP archives were gradually transformed following the introduction of the World Wide Web in 1993, and became full-fledged web sites as the Web grew more popular in 1994 and 1995. In 1997, the first commercial music download site appeared, MP3.com. Created by Michael Robertson, MP3.com was supported by advertising and offered free downloads of any of its thousands of songs.

Meanwhile, users needed ways to play these downloaded digital files. Many of the pre-MPEG file archives simply stored audio files in the uncompressed format used on the compact disc, or the Resource Interchange File Format announced by IBM and Microsoft in 1991. "Wave" files (identified by the .wav extension at the end of their file names) could be played by Apple Computer's Quicktime, introduced in 1991, or a number of other players. Fraunhofer researchers released a software-based MP3 player for the Microsoft Windows operating system, called Winplay, in 1995, but it was not as successful as Croatian software engineer Tomislav Uselac's AMP Engine, developed when he was a student at Zagreb University during and after 1996. After founding Advanced Multimedia Products (AMP) with Brian Litman to commercialize the player, Uselac reintroduced it as the AMP Playback Engine. Even more popular was Winamp, a shareware program similar to AMP created by former University of Utah students Justin Frankel and Dmitry Boldyrev, and distributed free over the Internet.

As the use of the Internet to access MP3 files grew, manufacturers began to introduce players that could accept the digital files from a personal computer, store them in memory, and play them back at the user's convenience. The first of these, the MPman, was succeeded by the more popular Rio player by Diamond Multimedia. Diamond was subsequently sued by the Recording Industry Association of America. In response to these threats,

GoodNoise, MP3.com, MusicMatch, Xing Technology, and Diamond Multimedia announced the formation of the MP3 Association, a group aimed at protecting the interests of companies associated with MP3 technology, in late 1998.

DIGITAL MILLENNIUM COPYRIGHT ACT

1999 was the year that MP3 music exploded in popularity. In part, this resulted from the high level of media coverage given to the Internet in the media through publications such as *Wired* magazine, which also contributed to an infamous stock market bubble. MP3.com became a publicly traded company that year, and on the first day that its stock opened for sale the price rose from an initial $28 to $72, despite the fact that the company had no source of revenue other than the sale of advertisements, which was not proving very lucrative. MP3.com's legal troubles would sap its resources, and the company would fail to find a way to make money while allowing free downloads of music.

The problems faced by online MP3 web sites, commercial or otherwise, were aggravated by the passage of the Digital Millennium Copyright Act of 1998, an international treaty that specifically banned the use or sale of "devices" (including software) intended to circumvent the copy-protection features of digital media files. This highly controversial act was attacked by critics as posing a threat to the free exchange of ideas and information, and for having the potential to stifle technological progress. Nonetheless, it was strongly supported by the powerful entertainment industry lobby and by organizations such as the RIAA.

Then in late 1999 Napster appeared. Napster was the name of a web site and a software package created by a young programmer named Shawn Fanning. When users registered with Napster, they could download software that created a special folder on the individual user's computer hard drive. In that folder, the user could store MP3 files for sharing. When the user was connected to the Internet, a central directory maintained on the Napster servers maintained an index of available songs. Users, by logging on at the Napster site and searching for titles of interest, could identify the locations of the songs they wanted and transfer them directly to their own computers. This new type of file transfer, called peer-to-peer sharing, apparently allowed Napster to skirt the laws, since it was the masses of users who were illegally copying songs, while Napster merely enabled them.

The RIAA nonetheless launched an immediate attack on Napster, filing a lawsuit against the company and pressuring U.S. universities (where music sharing was rampant) to ban the Napster site on computers used by students and faculty. In July 2000, the courts would demand that Napster prevent the downloading of copyrighted music via its service. The injunction would effectively shut Napster down, and although the RIAA abandoned its suit when the company was subsequently sold to a major publishing company, Napster's legal status was still uncertain. It went bankrupt in 2002.

Meanwhile the Nullsoft company, now the manufacturer of the Winamp MP3 player and a subsidiary of America Online, introduced a program called Gnutella, which offered a new type of peer-to-peer file sharing. Unlike Napster's software, Gnutella required no centralized database of song locations. Instead, users simply logged on to the Internet, launched the Gnutella software, and used a built-in search feature to look for songs; if the songs were available on other users' computers, a list of download locations would appear on the screen. Apparently realizing it had made a mistake, Nullsoft almost immediately stopped offering the Gnutella software, but since it had already made the source code available, other programmers could easily reshape it into new products. The result was a host of new, Gnutella-based file-sharing services, the most popular of which by 2002 was the Kazaa software of Sharman Networks, a British firm.

THE FUTURE OF RECORDING

Although it is impossible to predict at the time of this writing where these developments will lead, it is clear that another period of accelerated technological change is underway. These changes do not necessarily affect the nature of what people hear, nor are they making much of an impact on how or where they hear it. Studios are still producing the kinds of music they produced before, and people are still listening in their homes, in cars, and in public. But the use of digital recording, personal computers, and the Internet is already changing the patterns of the consumption of music. It is evident, particularly in the behavior of young people, that owning records and amassing collections are no longer as important to consumers as acquiring the music itself, represented by ephemeral and largely intangible digital files. Even more profound are the emerging changes in the recording industry, which is only gradually loosening its grip on the notion that its ultimate purpose is to manufacture something, rather than to distribute and promote music. The recording industry is making the transition from the manufacturing to the service sector of the economy, and in future years it will rely less on sales of physical media than on sales of songs.

Glossary

Cartridge (phonograph). *See* Pickup.

Dictaphone. The trade name of a wax-cylinder, sound recording and reproducing system based on the graphophone, and marketed as an office dictation machine from the 1880s to about 1950. Dictaphone became the generic name for dictation recorders made by other manufacturers and, after about 1940, using other recording media such as magnetic tape.

Gramophone. The trade name of the sound recording and reproducing system commercially introduced by Emile Berliner in 1894. The gramophone used a wax disc as its recording medium. The disc had to be electrochemically treated and copied before the discs could be played. The media used for the copies included forms of rubber, plastic, and lacquer. The gramophone became the basis of most consumer disc players in the twentieth century.

Graphophone. The sound recording and reproducing device patented by Chichester Bell and Charles Sumner Tainter in 1886. It consisted of a modified phonograph, using wax as the recording medium rather than tinfoil.

Magnetic recording. The method of recording sound as a continuously variable region of magnetism on a medium such as a steel wire, tape, or disc, or as a series of variably magnetized particles of a metallic compound.

Optical recording. Any of several methods for recording sound as a visible pattern on a medium. The earliest optical recorders responded directly to sound waves to "draw" the wave pattern on a medium such as paper. The motion picture industry in the 1930s began using photographic methods to record sound as a visible pattern at the edge of a movie print. In the late twentieth century, following the introduction of circuits to convert sound to binary data, optical recording was widely used to record sound as a series of pits or discolored areas on a metal or plastic medium to be scanned by a laser.

Phonautograph. The device invented by Frenchman Leon Scot in 1857, which recorded sound on a soot-coated disc, cylinder, or plate. The phonautograph was the first sound recorder, but it was not capable of sound reproduction.

Phonograph. The sound recording and reproducing device invented by Thomas Edison in 1877, which recorded sound in the form of indentations in a sheet of tinfoil. The term was also applied to Edison's later wax-cylinder and disc phonographs, and to inventions by others, particularly Emile Berliner's popular gramophone.

Pickup. An electromechanical device, used in conjunction with a phonograph record player, that converts the sound recording in the groove to an electrical current. Such a pickup can be piezoelectric or it can rely on electromagnetic transduction.

Piezoelectric. A class of crystal materials, including quartz, which respond to the application of electric pulses by vibrating. Piezoelectric materials also generate a minute voltage when pressed or twisted. Piezoelectrics sensitive enough to respond to the pressure of sound waves can be connected to an amplifier circuit so that the piezoelectric acts as a microphone. Attaching a stylus to certain piezoelectric materials produces an efficient form of phonograph pickup. During the 1920s and later, piezoelectric earphones were also commonly available.

Recording. The process of capturing sound waves in air and storing them on a medium, such as a disc, tape, motion picture film, or electronic memory device.

Reproduction. The process of converting a sound recording, stored on a medium such as a disc, tape, motion picture film, or electronic memory device, into audible sound.

Stereophonic recording. Usually refers to the practice, first demonstrated in the 1930s, of recording and later reproducing two separate sound recordings simultaneously. Technically, three or more simultaneous recordings is also stereophonic. If properly prepared, a stereophonic recording provides the listener with the illusion of physical depth, particularly in the horizontal plane.

Stylus. In the context of sound recording, a stylus usually refers to the engraving or embossing point that is held in contact with the recording medium to record or reproduce sounds.

Telegraphone. The trade name of a sound recording and reproducing device invented by Valdemar Poulsen around 1899. The telegraphone recorded sound magnetically on a steel wire, tape, or disc.

Transducer. In the context of sound recording, any device that converts sound to a flow of electricity, or a flow of electricity to sound. Transducers include microphones, loudspeakers, earphones, tape heads, and phonograph pickups.

Transistor. A replacement for the vacuum tube, invented in 1947. Unlike a vacuum tube, the transistor uses a solid crystal of a specially prepared material such as silicon or germanium. Electronic amplifiers and other systems employing transistors were often referred to as "solid state" to distinguish them from earlier vacuum tube designs.

Vacuum tube. The generic name for a device (also known as an electron tube) used in electronic amplification circuits. The vacuum tube is so named because it consists of a sealed glass or metal envelope, often tube-shaped, from which atmospheric gases have been removed. The tube typically contains one or more devices for amplifying electric currents.

Bibliography

Abbott, John E. "The Development of the Sound Film." *Journal of the Society of Motion Picture Engineers* 38 (1942): 541–548.

Aitken, Hugh. *The Continuous Wave: Technology and American Radio, 1900–1932*. Princeton, NJ: Princeton University Press, 1985.

———. *Syntony and Spark: The Origins of Radio*. Princeton, NJ: Princeton University Press, 1985.

Alexander, Robert Charles. *The Life and Works of Alan Dower Blumlein*. Oxford, England: Focal Press, 1999.

Austin, Mary. "Petrillo's War." *Journal of Popular Culture* 12 (summer 1978): 11–18.

Bachman, William. "From Transcription Disc to LP." *High Fidelity* 26 (April 1976): 58–60.

Begun, S. J. *Magnetic Recording*. New York: Rinehart & Company, 1949.

———. *Magnetic Recording: The Ups and Downs of a Pioneer*, ed. Mark Clark. New York: Audio Engineering Society, 2000.

Bennett, William R. "Secret Telephony as a Historical Example of Spread-Spectrum Communications." *IEEE Transactions on Communications*, vol. COM-31 (January 1983): 99.

Biel, Michael. "The Making and Use of Recordings in Broadcasting before 1936." Ph.D. diss., Northwestern University, 1977.

Boesen, Victor. *They Said It Couldn't Be Done: The Incredible Story of Bill Lear*. Garden City, NY: Doubleday and Company, 1971.

Bower, Tom. *The Paperclip Conspiracy: The Battle for the Spoils and Secrets of Nazi Germany*. London: Paladin Grafton Books, 1988.

Chanan, Michael. *Repeated Takes: A Short History of Recording and Its Effects on Music.* London: Verso, 1995.

Clark, Mark. "The Magnetic Recording Industry, 1878–1960." Ph.D. diss., University of Delaware, 1992.

Collins, Theresa M., et al., eds. *Thomas Edison and Modern America: A Brief History with Documents.* Boston: Bedford St. Martin's, 2002.

Cox, Arthur J., and Thomas Malim. *Ferracute: The History of an American Enterprise.* Bridgeton, NJ: Cowan Printing Co., 1985.

Crafton, Donald. *Talkies: American Cinema's Transition to Sound, 1926–1931.* Berkeley: University of California Press, 1999.

Daniel, Eric D., et al., eds. *Magnetic Recording: The First 100 Years.* Piscataway, NJ: IEEE Press, 1999.

Douglas, Susan. *Inventing American Broadcasting, 1899–1922.* Baltimore: Johns Hopkins University Press, 1987.

du Gay, Paul, et al. *Doing Cultural Studies: The Story of the Sony Walkman.* London: Sage Publications, 1997.

Dummer, G. W. A. *Electronic Inventions and Discoveries: Electronics from Its Earliest Beginnings to the Present Day*, 4th ed. Bristol, UK: Institute of Physics Publishing, 1997.

Dunlap, Orrin. *Radio's 100 Men of Science: Biographical Narratives of Pathfinders in Electronics and Television.* New York: Harper and Brothers, 1944.

Fagen, M. D. *A History of Engineering and Science in the Bell System: National Service in War and Peace.* N.p.: Bell Telephone Laboratories, 1978.

Fielding, Raymond, ed. *A Technological History of Motion Pictures and Television.* Berkeley: University of California Press, 1967.

Gillett, W. *The Phonograph and How to Construct It.* London: E. & F. N. Sponable, 1892.

Gimbel, John. *Science, Technology, and Reparations: Exploitation and Plunder in Postwar Germany.* Stanford, CA: Stanford University Press, 1990.

Goldmark, Peter, and Lee Edson. *Maverick Inventor: My Turbulent Years at CBS.* New York: Saturday Review Press, 1973.

Gomery, Douglas. "Failure and Success: Vocafilm and RCA Photophone Innovate Sound." *The Film Reader* 2 (1977): 213–221.

Harvith, John, and Susan Harvith. *Edison, the Musicians, and the Phonograph.* Westport, CT: Greenwood Press, 1987.

Heylin, Clinton. *Bootleg: The Secret History of the Other Recording Industry.* New York: St. Martin's Press, 1996.

Horning, Susan S. "Chasing Sound: The Culture and Technology of Recording Studios in America, 1877–1977." Ph.D. diss., Case Western Reserve University, 2002.

———. "From Polka to Punk: Growth of an Independent Recording Studio, 1934–1977." In *Music and Technology in the Twentieth Century*, edited by Hans-Joachim Braun, Baltimore: Johns Hopkins University Press, 2002. 136–147.

Hughes, Thomas P. *American genesis: A Century of Invention and Technological Enthusiasm.* New York: Viking, 1989.

Immink, Kees A. Schouhamer. "The Compact Disc Story." *Journal of the Audio Engineering Society* 46 (May 1998): 458–460.

International Federation of Phonograph Industries. "Regional Summaries." 1999. http://www.ifpi.org/site-content/statistics/regional_summaries.html.

Israel, Paul. *Edison: A Life of Invention*. New York: John Wiley and Sons, 1998.

Israel, Paul, et al., eds. *The Papers of Thomas A. Edison*, vol. 4. Baltimore: Johns Hopkins University Press, 1998.

Johnson, E. R. Fenimore. *His Master's Voice Was Eldridge R. Johnson*. N.p.: privately published, 1974.

Jones, Geoffrey. "The Gramophone Company: An Anglo-American Multinational, 1898–1931." *Business History Review* 59 (spring 1985): 76–100.

Jones, Steven. *Rock Formation: Music, Technology, and Mass Communication*. New York: Sage Publications, 1992.

Kenney, William H. *Recorded Music in American Life: The Phonograph and Popular Memory, 1890–1945*. New York: Oxford University Press, 1999.

Kraft, James P. *Stage to Studio: Musicians and the Sound Revolution, 1890–1950*. Baltimore: Johns Hopkins University Press, 1996.

Lafferty, William. "The Blattnerphone: An Early Attempt to Introduce Magnetic Recording into the Film Industry." *Cinema Journal* 22 (summer 1983): 18–37.

———. "The Early Development of Magnetic Sound Recording in Broadcasting and Motion Pictures, 1928–1950." Ph.D. diss., Northwestern University, 1981.

Leiter, Robert D. *The Musicians and Petrillo*. New York: Bookman and Associates, 1953.

Magoun, Alexander B. "Shaping the Sound of Music: The Evolution of the Phonograph Record, 1877–1950." Ph.D. diss., University of Maryland, 2000.

Martin, George. *With a Little Help from My Friends: The Making of Sgt. Pepper.* Boston: Little, Brown, 1994.

Martland, Peter. *EMI: The First 100 Years*. London: EMI Group, 1997.

McGinn, Robert. "Stokowski and the Bell Telephone Laboratories: Collaboration in the Development of High-Fidelity Sound Reproduction." *Technology and Culture* 24 (January 1983): 38–75.

Millard, Andre. *America on Record: A History of Recorded Sound*. London: Cambridge University Press, 1995.

———. *Edison and the Business of Innovation*. Baltimore: Johns Hopkins University Press, 1990.

Mooney, Mark. "The History of Magnetic Recording." *Hi-Fi Tape Recording* 5 (February 1958): 31–37.

Morton, David. "Armour Research Foundation and the Wire Recorder: How Academic Entrepreneurs Fail." *Technology and Culture* 39 (1998): 213–244.

———. "Disc Television Recording." In *The Encyclopedia of 20th Century Technology*, edited by Colin Hempstead and William E. Worthington. London: Routledge, forthcoming.

———. *A History of Electronic Entertainment since 1945*. Piscataway, NJ: IEEE, Inc., 1999.

———. "The History of Magnetic Recording in the United States, 1888–1978." Ph.D. diss., Georgia Institute of Technology, 1995.

———. *Off the Record: The Technology and Culture of Sound Recording in America*. New Brunswick, NJ: Rutgers University Press, 2000.

———. " 'The Rusty Ribbon': John Herbert Orr and the Making of the Magnetic Recording Industry, 1945–1960." *Business History Review* 67 (Winter 1993): 589–622.

Mullin, John T. "Creating the Craft of Tape Recording." *High Fidelity* (April 1976): 62–67.

Musser, Charles. *Thomas Edison and His Kinetographic Motion Pictures*. New Brunswick, NJ: Friends of the Edison National Historic Site, 1995.

Nebeker, Frederick. *Signal Processing: The Emergence of a Discipline*. New Brunswick, NJ: IEEE History Center, 1998.

Rashke, Richard L. *Stormy Genius: The Life of Aviation's Maverick, Bill Lear*. Boston: Houghton Miflin Co., 1985.

Read, Oliver, and Walter L. Welch. *From Tin Foil to Stereo: Evolution of the Phonograph*, 2nd ed. Indianapolis, IN: Howard W. Sams & Co., 1959, 1976.

Rosenberg, Robert A., et al., eds. *The Papers of Thomas A. Edison*, vol. 3 Baltimore: Johns Hopkins University Press, 1994.

Sanjek, Russell. *American Popular Music and Its Business: The First Four Hundred Years*, vol. 3. New York: Oxford University Press, 1988.

Schiffer, Michael Brian. *The Portable Radio in American Life*. Tucson: University of Arizona Press, 1991.

Schmidt-Horning, Susan. "Chasing Sound: The Culture and Technology of Recording Studios in America, 1877–1977." Ph.D. diss., Case Western Reserve University, 2002.

Schwartz, David. "Strange Fixation: Bootleg Sound Recordings Enjoy the Benefits of Improving Technology." *Federal Communications Law Journal* 47 (1994–1995). May 18, 2004, http://www.law.indiana.edu/fclj.

Smith, Oberlin. "Some Possible Forms of the Phonograph." *Electrical World* 8 (September 1888): 116.

Smits, F. M., ed. *A History of Engineering and Science in the Bell System: Electronics Technology (1925–1975)*. N.p.: AT&T Bell Laboratories, 1985.

"Sound Picture Statistics for 1930." *Electronics* 2 (March 1931): 538–539.

Thompson, Emily. "Machines, Music, and the Quest for Fidelity: Marketing the Edison Phonograph in America, 1877–1925." *The Musical Quarterly* 79 (1995): 131–171.

Wallerstein, Edward. "Creating the LP Record." *High Fidelity* 26 (April 1976): 56–61.

Weis, Elisabeth, and John Belton, eds. *Film Sound: Theory and Practice*. New York: Columbia University Press, 1985

Welch, Walter, et al. *From Tin Foil to Stereo: The Acoustic Years of the Phonograph Industry, 1877–1929*. Gainesville: University Press of Florida, 1994.

Yates, JoAnne. *Control through Communication: The Rise of System in American Management*. Baltimore: Johns Hopkins University Press, 1989.

ONLINE RESOURCES

Edison National Historic Site (National Park Service)
http://www.nps.gov/edis/home.htm

History of the MP3 Standard
http://www.iis.fraunhofer.de/amm/techinf/layer3/index.html

Recording Technology History
http://history.sandiego.edu/gen/recording/notes.html

Recording-History.org
http://www.Recording-History.org

Thomas A. Edison Papers Project at Rutgers University
http://edison.rutgers.edu/taep.htm

U.S. Library of Congress Recorded Sound Reference Center
http://www.loc.gov/rr/record/

Index

About the Author

DAVID L. MORTON JR. is a historian of technology with expertise in the history of sound recording, electronics, and electric power. He is the former Research Historian for the Institute of Electrical and Electronics Engineers.